T0140026

Studies in Classification, Data Analysis, and Knowledge Organization

More information about this series at http://www.springer.com/series/1564

Francesca Greselin • Laura Deldossi •
Luca Bagnato • Maurizio Vichi
Editors

Statistical Learning of Complex Data

Editors

Francesca Greselin
Department of Statistics and Quantitative Methods
University of Milano-Bicocca
Milan, Italy

Laura Deldossi
Department of Statistical Sciences
Università Cattolica del Sacro Cuore
Milan, Italy

Luca Bagnato
Department of Economic and Social Sciences
Università Cattolica del Sacro Cuore
Piacenza, Italy

Maurizio Vichi
Department of Statistical Sciences
Sapienza University of Rome
Rome, Italy

ISSN 1431-8814 ISSN 2198-3321 (electronic)
Studies in Classification, Data Analysis, and Knowledge Organization
ISBN 978-3-030-21139-4 ISBN 978-3-030-21140-0 (eBook)
https://doi.org/10.1007/978-3-030-21140-0

Mathematics Subject Classification (2010): 62-06, 62-07, 62Fxx, 62Gxx, 62Hxx, 62Jxx, 62Kxx

This Springer imprint is published by the registered company Springer Nature Switzerland AG.
The registered company address is: Gewerbestrasse 11, 6330 Cham, Switzerland

Preface

This volume collects selected papers presented at the 11th biannual meeting of the Classification and Data Analysis Group (CLADAG) of the *Società Italiana di Statistica*, held in Milan, September 13–15, 2017.

The program of the conference included 142 presentations, organized in 3 plenary talks, 21 invited sessions, 18 contributed sessions, and a poster session. We wish to express our gratitude to the authors, whose enthusiastic participation made the meeting possible. The conference provided a vibrant international forum for discussion and a mutual exchange of knowledge, thanks to the 163 attendees and authors coming from several European countries, like Austria, Denmark, France, Germany, Great Britain, Ireland, the Netherlands, Norway, Poland, Spain, and Switzerland, as well as from the United States and Japan. The Scientific Committee of the Conference was chaired by Francesco Mola, and Francesca Greselin was the Chairperson of the Local Organizing Committee.

The topics of Plenary and Invited Sessions were carefully chosen by the Scientific Committee in view of the CLADAG mission: to promote methodological, computational, and applied research in classification, data analysis, and multivariate statistics. We thank all the organizers of the sessions for inviting renowned speakers. We extend our gratitude to all the chairpersons and active participants, whose interesting comments and suggestions made the conference a real motivating event.

The 20 manuscripts included in the present volume were selected, through a blind review process, among the ones presented at the conference and later submitted for the publication in the Springer book series. We are greatly indebted to the referees (at least two scholars were involved for each paper) for the time and effort they spent in such a careful review.

The volume is divided into five parts as follows: Clustering and Classification, Exploratory Data Analysis, Statistical Modeling, Graphical Models, and Big Data Analysis.

The first part, Clustering and Classification, contains methodologically oriented papers. The paper by Fordellone and Vichi presents the combined usage of unsupervised classification with supervised methods to enhance the assessment and the interpretation of the obtained partition; Rainey, Tortora, and Palumbo are the authors of the second work, that introduces a parametric version of probabilistic distance clustering based on the Gaussian and the Student's t multivariate density distributions. Cappozzo and Greselin deal with an interesting application to wine authenticity studies based on robust clustering methods, where a mixture of Gaussian factors is used to ascertain varietal genuineness and distinguish potentially doctored food. Simulation results are presented by Alfò, Nieddu, and Vitiello to study the performance of the cluster-weighted beta regression in a variety of empirical settings. The results show that the model captures individual specific unobserved heterogeneity and its link with observed covariates and indicate some shortcomings related to the scale of the observed quantities. The last paper of this part, by Ranalli and Rocci, presents an overview on a recent model-based approach to cluster ordinal variables. The aim is to extend the proposal to the case where noise dimensions or variables are present, and to generalize the model to mixed-type data.

The second part, devoted to Exploratory Data Analysis, contains a first paper by Bove, Ruta, and Mastandrea where multidimensional scaling and unfolding allow to easily detect preference order, size of asymmetry, and relationships between subjects and stimuli coming from the curvature of architectural facades. In the second paper, De Stefano, Vitale, and Zaccarin submit exploratory results obtained by adopting two well-known community detection methods and a new proposal, aiming at discovering groups of scientists in the coauthorship network of Italian academic statisticians. In their study, Okada and Tsurumi extend the asymmetric multidimensional scaling to analyze differences among consumers and to show how each consumer or a group of consumers relates to brands in brand switching. In the fourth paper, Solaro faces the problem of comparing the results obtained with different imputation methods, so critical in many fields of application, such as cardiovascular studies in the medical context. She also assesses the quality of imputation, through the dissimilarity profile analysis.

The third part refers to Statistical Modeling. The first paper is by Altimari, Balzano, and Zezza who introduce an extended version of the Economic Vulnerability Index, adopted by the United Nations. By the partial least square approach to structural equation model, an estimate of the extended index is obtained using data from a panel of 98 countries over 19 years. Ascari, Migliorati, and Ongaro present two Bayesian procedures—both based on Gibbs sampling—to estimate the parameters of the flexible Dirichlet. This distribution is particularly suited for compositional data, thanks to its mixture structure and to the additional parameters that allow for a more flexible modeling of the covariance matrix. Fiori and Motta contribute a new stochastic model of firm dynamics that leads to a Dagum distribution for the size of business firms operating in a given industry. The model relies on a stochastic growth process that was originally introduced in the context of income inequality studies and sheds new light upon the connections between growth dynamics and the meaning of parameters that appear in the steady-state distribution of firm size. A mixture of Mincer's models with concomitant variables is proposed by Mazza, Battisti, Ingrassia, and Punzo. The new model provides a flexible generalization of the Mincer model, a breakdown of the population into several homogeneous subpopulations, and an explanation of the unobserved heterogeneity. The proposal is motivated and illustrated via an application to data provided by the Bank of Italy's Survey of Household Income and Wealth in 2012. In his paper, Nakai presents interesting results about the impact of women's own human capital on contribution to household income, by using multinomial logistic regression upon data coming from two national social surveys conducted in 1985, 1995, 2005, and 2015 in Japan. The last paper of this part by Vernizzi and Nakai introduces an optimization algorithm that improves the size of the final dataset after applying the listwise deletion method. The proposed weighted optimization method has been applied, first to some toy examples, and then to the National Longitudinal Survey of Youth dataset.

The fourth part—devoted to Graphical Models—contains a first paper by Marella, Vicard, Vitale, and Ababei, which proposes a procedure based on non-parametric Bayesian networks to detect and correct measurement error. The novel procedure is evaluated on a validation sample associated to the Bank of Italy survey on household income and wealth. In the second paper, Musella, Vicard, and Vitale address the issue of Bayesian network structural learning for non-paranormal data. They propose a modified version of the Copula Grow-Shrink algorithm whose high performance is proved through a simulation study. In addition, an application to Italian energy market data is also provided. The third and last paper of this part is by Nicolussi and Cazzaro; it aims to incorporate the contest-specific independence conditions in graphical models. The authors take advantage of the Hierarchical Multinomial Marginal parametrization to represent the dependence relationships. The proposal is applied on the study of the trend of innovation degree for Italian enterprises.

The last part is devoted to Big Data Analysis. The first paper, by Giuffrida, Gozzo, Rinaldi, and Tomaselli, proposes a new approach to address, in the Big Data world, the analysis of relational structures to improve actionable analytics-driven decision patterns. An application to model online news is provided. In the second paper, Pesce, Riccomagno, and Wynn discuss some issues related to the use of experimental design to help establish causation in complex models. The question to remove bias is also considered: various solutions are discussed, including randomization.

We cannot conclude this brief introduction without some further thanks. We gratefully acknowledge the Department of Statistics and Quantitative Methods and the University of Milano Bicocca, which strongly supported the CLADAG conference. We shared constructing ideas with some colleagues from Milan, valuably supported by their Institutions: Piercesare Secchi from Politecnico, Laura Deldossi from the Università Cattolica del Sacro Cuore, Pieralda Ferrari from the Università di Milano, and Raffaella Piccarreta from the Università Bocconi. To all of them go our most sincere appreciations. Special thanks are due to the members of the Local Organizing Committee. They all did a great job! A particular mention goes to Mariangela Zenga for her tireless activity and enthusiasm.

We are grateful to Oracle Corporation, Data Reply, and Fondazione Cariplo. They supported the event and made possible the data challenge for Young Cladag. They also sponsored the final concert. Maestro Alessandro Arnoldo conducted the Milan Chamber Orchestra, and created a delightful moment for all participants.

Finally, we acknowledge Dr. Veronika Rosteck of Springer-Verlag, Heidelberg, for her support and dedication to the making of this volume.

Milan, Italy — Francesca Greselin
Milan, Italy — Laura Deldossi
Piacenza, Italy — Luca Bagnato
Rome, Italy — Maurizio Vichi
December 2018

Contents

Contributors

Dan Ababei LightTwist Software, Delft, Netherlands

Marco Alfò Dipartimento di Scienze Statistiche, Sapienza University of Rome, Roma, Italy

Ambra Altimari Department of Economics and Law, University of Cassino and Southern Lazio, Cassino, Italy

Roberto Ascari Department of Economics, Management and Statistics, University of Milano-Bicocca, Milano, Italy

Simona Balzano Department of Economics and Law, University of Cassino and Southern Lazio, Cassino, Italy

Michele Battisti Dipartimento di Scienze Giuridiche, della Società e dello Sport, University of Palermo, Palermo, Italy

Giuseppe Bove Department of Education, Roma Tre University, Roma, Italy

Andrea Cappozzo Department of Statistics and Quantitative Methods, University of Milano-Bicocca, Milano, Italy

Manuela Cazzaro Department of Statistics and Quantitative Methods, University of Milano-Bicocca, Milano, Italy

Domenico De Stefano Department of Social and Political Sciences, University of Trieste, Trieste, Italy

Anna Maria Fiori Department of Statistics and Quantitative Methods, University of Milano-Bicocca, Milano, Italy

Mario Fordellone Dipartimento di Scienze Statistiche, Sapienza University of Rome, Roma, Italy

Giovanni Giuffrida Department of Political and Social Sciences, University of Catania, Catania, Italy

Simona Gozzo Department of Political and Social Sciences, University of Catania, Catania, Italy

Francesca Greselin Department of Statistics and Quantitative Methods, University of Milano-Bicocca, Milano, Italy

Salvatore Ingrassia Dipartimento di Economia e Impresa, University of Catania, Catania, Italy

Daniela Marella Dipartimento di Scienze della Formazione, Roma Tre University, Roma, Italy

Stefano Mastandrea Department of Education, Roma Tre University, Roma, Italy

Angelo Mazza Dipartimento di Economia e Impresa, University of Catania, Catania, Italy

Sonia Migliorati Department of Economics, Management and Statistics, University of Milano-Bicocca, Milano, Italy

Anna Motta Il Sole 24 Ore, Milano, Italy

Flaminia Musella Department of Research, Link Campus University, Roma, Italy

Miki Nakai Department of Social Sciences, College of Social Sciences, Ritsumeikan University, Kyoto, Japan

Federica Nicolussi Department of Economics, Management and Quantitative Methods, University of Milano, Milano, Italy

Luciano Nieddu Università degli Studi Internazionali di Roma, Roma, Italy

Akinori Okada Research Institute, Tama University, Tokyo, Japan

Andrea Ongaro Department of Economics, Management and Statistics, University of Milano-Bicocca, Milano, Italy

Francesco Palumbo Department of Political Sciences, University of Naples Federico II, Napoli, Italy

Elena Pesce Department of Mathematics, University of Genoa, Genova, Italy

Antonio Punzo Dipartimento di Economia e Impresa, University of Catania, Catania, Italy

Christopher Rainey Department of Mathematics and Statistics, San José State University One Washington Square, San Jose, CA, USA

Monia Ranalli Department of Economics and Finance , University of Tor Vergata, Roma, Italy

Eva Riccomagno Department of Mathematics, University of Genoa, Genova, Italy

Francesco Mazzeo Rinaldi Department of Political and Social Sciences, University of Catania, Catania, Italy

Roberto Rocci Department of Economics and Finance, University of Tor Vergata, Roma, Italy

Nicole Ruta Cardiff School of Art & Design, Cardiff, UK

Nadia Solaro Department of Statistics and Quantitative Methods, University of Milano-Bicocca, Milano, Italy

Venera Tomaselli Department of Political and Social Sciences, University of Catania, Catania, Italy

Cristina Tortora Department of Mathematics and Statistics, San José State University, One Washington Square, San Jose, CA, USA

Hiroyuki Tsurumi Yokohama National University, Yokohama-shi, Japan

Graziano Vernizzi Department of Physics and Astronomy, Siena College, Loudonville, NY, USA

Paola Vicard Dipartimento di Economia, Roma Tre University, Roma, Italy

Maurizio Vichi Department of Statistical Sciences, Sapienza University of Rome, Rome, Italy

Maria Prosperina Vitale Department of Political and Social Studies, University of Salerno, Fisciano, Italy

Vincenzina Vitale Dipartimento di Scienze Sociali ed Economiche, Sapienza University of Rome, Roma, Italy

Cecilia Vitiello Dipartimento di Scienze Statistiche, Sapienza University of Rome, Roma, Italy

Henry P. Wynn Department of Statistics, London School of Economics, London, UK

Susanna Zaccarin Department of Business, Economic, Mathematics and Statistics, University of Trieste, Trieste, Italy

Gennaro Zezza Department of Economics and Law, University of Cassino and Southern Lazio, Cassino, Italy

Part I
Clustering and Classification

Cluster Weighted Beta Regression: A Simulation Study

Marco Alfó, Luciano Nieddu, and Cecilia Vitiello

Abstract In several application fields, we have to model a response that takes values in a limited range. When these values may be transformed into rates, proportions, concentrations, that is to continuous values in the unit interval, beta regression may be the appropriate choice. In the presence of unobserved heterogeneity, for example when the population of interest is composed by different subgroups, finite mixture of beta regression models could be useful. When conditions of exogeneity of the covariates set are not met, extended modeling approaches should be proposed. For this purpose, we discuss the class of cluster-weighted beta regression models.

Keywords Beta regression · Finite mixtures · Cluster-weighted regression

1 Introduction

Frequently, we are interested in describing the (conditional on covariates) distribution of a continuous response variable taking values in a limited interval, which could be mapped onto the open unit interval. Such variables can be observed in a wide variety of empirical situations, in the form of relative frequencies of an event (e.g., number of votes obtained by a candidate out of the total votes cast at an election), fractions of a continuous variable (e.g., amount of GDP due to a specific economic sector), performance ratings (e.g. student performances when compared to the maximum performance attainable), and limited-valued indexes (e.g., the relative Gini index). Examples of this framework can be found in medical (see, e.g., [12, 21, 22]), education (see, e.g., [3]), and economics [2, 6] research.

M. Alfó · C. Vitiello
Dipartimento di Scienze Statistiche, Sapienza University of Rome, Rome, Italy
e-mail: marco.alfo@uniroma1.it; cecilia.vitiello@uniroma1.it

L. Nieddu (✉)
UNINT, Rome, Italy
e-mail: l.nieddu@unint.eu

F. Greselin et al. (eds.), *Statistical Learning of Complex Data*,
Studies in Classification, Data Analysis, and Knowledge Organization,
https://doi.org/10.1007/978-3-030-21140-0_1

In this context, the use of a standard linear model is not a feasible solution (see, e.g., [17] and [14]). A naive solution is to map the response onto the real line (for example using a probit transform), so that a standard regression model could be used [7]. While this approach is preferable when compared to a linear model for the original response variable, it suffers from the well-known shortcomings, see [1] and [19]. A viable alternative is to use a regression model based on a conditional beta distribution for the response Y given the p-dimensional vector of covariates $\mathbf{x}$, that is $Y \mid \mathbf{x} \sim \mathscr{B}(p, q), \quad Y \in (0, 1)$, with parameters $p, q > 0$. This model has been introduced by Ferrari and Cribari-Neto [8], and extended by Ospina and Ferrari [15, 16] to account for those cases where the response is defined over the closed interval $[0, 1]$, via (factorizable) mixtures of discrete and continuous distributions. For modeling purposes, Ferrari and Cribari-Neto [8] proposed the following parameterization:

$$\mathbb{E}(Y) = \mu = \frac{p}{p+q}; \quad Var(Y) = \frac{\mu(1-\mu)}{1+\phi} \tag{1}$$

where $\phi = p + q$ represents the *precision* parameter. The beta distribution may assume different shapes according to different combinations of the (p, q), or (μ, ϕ), parameters. Parameter estimates can be obtained using a maximum likelihood (ML) approach, see [8]; for this purpose, the `R` library `Betareg`, [5], can be used. In some cases, however, individual heterogeneity is only partially accounted for by the observed covariates, and continuous/discrete mixed effect beta regression should be taken into consideration. In particular, when the omitted covariates can be described by a latent variable with a discrete distribution, or the population of interest is composed by several subgroups, characterized by different values of regression parameters, finite mixtures of beta regressions represent a viable option. This model has been introduced in the literature by Grün et al. [10], who designed the `R` library `betamix`. As in standard finite mixture models, this is based on the (sometimes non-explicitly stated) assumption of assignment independence, see [11], which can be also considered as a sort of *exogeneity of observed covariates* with respect to the discrete latent variable. Two issues are, in this case, of interest. First, due to the wide range of different shapes the beta distribution may take, a further parameterization has been proposed by Chen [4], see also [1]. This is motivated by identifiability issues and it is based on the subclass of unimodal beta densities. This prevents the problem of a U-shaped distribution being fitted by a mixture of two J-shaped ones. However, in the present context, we prefer an approach based on modeling the mean of the response, rather than its mode. Second, in several empirical cases, the omitted covariates described by the latent variables and the observed covariates may not be independent, and this may question the reliability of parameter estimates. In fact, if we do not account for such a dependence, the estimated impact of observed covariates may be either due to a direct effect on the response or to the (mediated) effect of omitted covariates. In a former paper, we introduced the class of cluster-weighted beta regression models, see [9], to capture individual specific unobserved heterogeneity and its link with observed covariates (see [13] and [18]).

In the next section, after discussing the standard finite mixture approach, we motivate the proposed approach. The EM algorithm for ML parameter estimation is sketched in Sect. 3 while, in Sect. 4, we report the results of a simulation study. Some concluding remarks are drawn in Sect. 5.

2 The Model

Let $(Y, \mathbf{X})$ be the set including a response variable Y and a covariates vector $\mathbf{X}$; let the corresponding population be partitioned into K subpopulations, referred to as components, and let $\pi_k(\mathbf{x}_i)$ denote the prior probability that unit i belongs to component $k = 1, \dots, K$. We associate to subpopulations the indicator vector with elements $z_{ik} = 1$ if unit i belongs to component k. We further assume that, conditional on being in the k-th component, $k = 1, \dots, K$, the following model holds

$$Y_i \mid X_i, z_{ik} = 1 \sim \mathscr{B}(p_{ik}, q_{ik}), \quad p_{ik}, q_{ik} > 0 \tag{2}$$

$$\mu_{ik} = \frac{p_{ik}}{p_{ik} + q_{ik}} = \eta_{ik} = h_1\left(\mathbf{x}_i' \boldsymbol{\beta}_k\right) \qquad \phi_{ik} = \frac{\mu_{ik}(1 - \mu_{ik})}{1 + \phi_{ik}} = \xi_{ik} = h_2\left(\mathbf{x}_i' \boldsymbol{\gamma}_k\right).$$

That is, each component is associated to possibly varying parameter vectors, $\boldsymbol{\beta}_k$ and $\boldsymbol{\gamma}_k$, in the models for the mean and the precision parameter of that component. However, the component indicators are not observed and, for purpose of estimation in an ML framework, we need to define the observed data joint density:

$$f(y_i \mid \mathbf{x}_i) = \sum_{k=1}^{K} f(y_i \mid \mathbf{x}_i, z_{ik} = 1)\pi_k(\mathbf{x}_i) \quad . \tag{3}$$

While the first term in the sum denotes the (conditional) beta density, the term $\pi_k(\mathbf{x}_i)$, $k = 1, \dots, K$, has been previously defined. A usual assumption is that $\pi_k(\mathbf{x}_i) = \pi_k \; \forall k$, known as *assignment independence*. When the independence is not met, this assumption may lead to a severe bias in model parameter estimates. This could be simply motivated by looking at the graph in Fig. 1; here, $\mathbf{x}_i$ has a non-zero impact on Y_i either directly or indirectly through z_{ik}. By adopting an independence assumption, we use a misspecified model. As a result, we remove the gray dashed edge from $\mathbf{x}_i$ to z_{ik}, and inflate the impact of the covariates on the response Y_i. In the case $\mathbf{x}_i$ and z_{ik} are not independent, we need to either model the impact of covariates on the component indicator, or go for the specification of the *marginal* density of the couple $(Y, \mathbf{X})$. In the former case, we cannot distinguish between the *direct* and the *indirect* impact of $\mathbf{X}$; in fact, there is no way, as opposed to the case when repeated observations are available, to test whether the effect of the observed covariates on Y or on Y through Z. For the latter, we may turn to consider a more flexible family of mixture models, the cluster-weighted models (CWM), which can

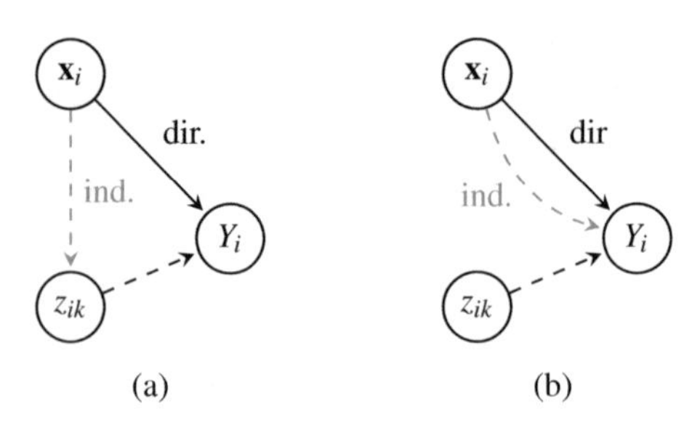

Fig. 1 A simple path diagram. (**a**) *True* model. (**b**) Misspecified model

be obtained by releasing the assignment independence hypothesis. This approach has been introduced by Gershenfeld [9] as a model based clustering approach for the couple $(Y, \mathbf{X})$. See [13] and [18] for extensions. The model is:

$$f(y_i, \mathbf{x}_i) = \sum_{k=1}^{K} g(\mathbf{x}_i \mid z_{ik} = 1) f(y_i \mid \mathbf{x}_i, z_{ik} = 1)\pi_k \quad . \tag{4}$$

Here, $f(y_i \mid \mathbf{x}_i, z_{ik} = 1)$ is the density of the response conditional on the set of covariates and the component the unit belongs to, and $g(\mathbf{x}_i \mid z_{ik} = 1)$ denotes the distribution of the covariates in the specific component $k = 1, \ldots, K$.

3 ML Parameter Estimation

When the covariates are continuous, it is customary to adopt a component-specific multivariate Gaussian density, that is $\mathbf{X} \mid z_{ik} = 1 \sim MVN(\boldsymbol{\nu}_k, \boldsymbol{\Sigma}_k)$. With non-continuous or mixed-type covariates more general models should be used to describe component-specific covariates distribution, [13]. A (conditional) beta model:

$$g_1(\mu_i) = \eta_k = \mathbf{x}_i' \beta_k \qquad g_2(\phi_i) = \xi_k = \mathbf{x}_i' \gamma_k \tag{5}$$

is used for the response, while the link functions $g_1(\cdot)$ and $g_2(\cdot)$ are monotone and twice differentiable. Given these modeling assumptions, the observed data likelihood is:

$$L(\psi) = \prod_{i=1}^{n} \sum_{k=1}^{K} f(y_i \mid \mathbf{x}_i, \boldsymbol{\beta}_k, \boldsymbol{\gamma}_k) g(\mathbf{x}_i \mid \boldsymbol{\upsilon}_k, \boldsymbol{\Sigma}_k)\pi_k$$

where $\boldsymbol{\theta}_k = (\boldsymbol{\beta}_k, \boldsymbol{\gamma}_k)$ denotes the vector of regression parameters, while $\boldsymbol{\omega}_k = (\boldsymbol{\nu}_k, \boldsymbol{\Sigma}_k)$, $k = 1, \ldots, K$, represents the parameters for the conditional Gaussian distribution for the set of observed covariates. Last, $\boldsymbol{\psi} = (\theta_1, \ldots, \theta_K, \omega_1, \ldots, \omega_K, \pi_1, \ldots, \pi_{K-1})$ denotes the global set of model parameters. Let $\ell_c(\boldsymbol{\psi})$ denote the log-likelihood function for the complete data $\{y_i, \mathbf{x}_i, \mathbf{z}_i\}_{i=1,\ldots,n}$. For fixed K, at the r-th iteration of the EM algorithm, $r = 1, \ldots,$ the E-step computes the expected value of the log-likelihood function for complete data, conditional on the observed data and the current parameters estimates $\boldsymbol{\psi}^{(r-1)}$. This is referred to as $Q(\boldsymbol{\psi} \mid \boldsymbol{\psi}^{(r-1)})$. In the M-step of the algorithm, this function is maximized with respect to ψ. The algorithm alternates the two steps until convergence, defined in terms of the norm of the difference between two subsequent values of model parameters:

E-step Compute $Q(\boldsymbol{\psi} \mid \boldsymbol{\psi}^{(r-1)}) = \mathbb{E}_{\psi^{(r-1)}}(\ell_c(\psi) \mid \mathbf{X}, y)$
M-step maximize $Q(\boldsymbol{\psi} \mid \boldsymbol{\psi}^{(r-1)})$ w.r.t. $\boldsymbol{\psi}$ to obtain updated estimates $\psi^{(r)}$

In the E-step the missing component indicator z_{ik} is replaced by the corresponding conditional expectation

$$w_{ik}^{(r)} = \frac{f(y_i \mid \mathbf{x}_i, \theta_k^{(r-1)}) g(\mathbf{x}_i \mid \omega_k^{(r-1)}) \pi_k^{(r-1)}}{\Sigma_{l=1}^{K} f(y_i \mid \mathbf{x}_i, \theta_l^{(r-1)}) g(\mathbf{x}_i \mid \omega_l^{(r-1)}) \pi_l^{(r-1)}},$$

that represents the posterior probability that the i-th unit comes from the k-th component, $i = 1, \ldots, n$, $k = 1, \ldots, K$. ML equations for the beta model and the Gaussian density parameters are given by the following expressions

$$\frac{\partial \ell(\cdot)}{\partial \theta_k} = \sum_{i=1}^{n} w_{ik}^{(r)} \frac{\partial \log(f(y_i \mid \mathbf{x}_i, \theta_k))}{\partial \theta_k} = 0 \quad \text{and} \quad \frac{\partial \ell(\cdot)}{\partial \omega_k} = \sum_{i=1}^{n} w_{ik}^{(r)} \frac{\partial \log(g(\mathbf{x}_i \mid \omega_k))}{\delta \omega_k} = 0.$$

Both expressions are weighted score equations with weights given by $w_{ik}^{(r)}$. The updated estimates for the Gaussian density parameters are available in closed form:

$$\hat{\nu}_k^{(r)} = \frac{\sum_{i=1}^{n} \mathbf{x}_i w_{ik}^{(r)}}{\sum_{i=1}^{n} w_{ik}^{(r)}} \quad \text{and} \quad \hat{\Sigma}_k^{(r)} = \frac{\sum_{i=1}^{n} w_{ik}^{(r)} (\mathbf{x}_i - \hat{\nu}_k^{(r)})(\mathbf{x}_i - \hat{\nu}_k^{(r)})'}{\Sigma_{i=1}^{n} w_{ik}^{(r)}}.$$

while $\hat{\pi}_k^{(r)} = \sum_{i=1}^{n} w_{ik}^{(r)}/n$, a well known result in finite mixtures.

4 Simulation Study

To study the proposed model performance in a variety of empirical settings, we defined the following simulation experiment. The aim is to evaluate the model's ability to recover the *true* parameter values and the coverage of the corresponding confidence intervals. Further, we evaluate whether the estimated component membership recover the *true* partition of the data. To make the data easy to interpret and visualize, the covariates $X_i = (X_{i1}, X_{i2})$ have been drawn from a multivariate

Gaussian density $\mathbf{X}_i|z_{ik} = 1 \sim MVN(\mu_k, \Sigma)$ and the number of groups has been set to $K = 3$ where $\mu_k = \begin{cases} \{0 \quad 0\} & k = 1 \\ \{-1 \; +1\} & k = 2 \\ \{+1 \; +2\} & k = 3 \end{cases}$

For each component we have drawn $n_k = 700$, $k = 1, 2, 3$ units resulting in a total sample size n =2100. Two possible scenarios have been considered for the covariance matrix, namely:

$$\text{A: } \Sigma = \begin{bmatrix} \sigma^2 & 0 \\ 0 & \sigma^2 \end{bmatrix}; \quad \sigma^2 = \{0.09 + a \cdot 0.02\}, \quad a = 1, \dots 10$$

$$\text{B: } \Sigma = b \begin{bmatrix} \sigma_1^2 & \rho\sigma_1\sigma_2 \\ \rho\sigma_1\sigma_2 & \sigma_2^2 \end{bmatrix}; \quad \sigma_1 \in \{1, 2\}, \sigma_2 \in \{1, 2\}, \rho \in \{0, 0.2, 0.8\}$$

where b is a multiplicative factor assuming values in $\{0.09, 0.11, 0.13\}$. Values for the response variable $Y_i|z_{ik} = 1$ have been drawn from a beta distribution with parameters (μ_{ik}, ϕ_{ik}) which have been chosen, conditionally on the component membership, according to the linear predictors

$$\mu_{ik} = \beta^*_{0k} + \beta^*_{1k}x_{i,1} + \beta^*_{2k}x_{i,2} \qquad \phi_{ik} = \gamma^*_{0k} + \gamma^*_{1k}x_{i,1} + \gamma^*_{2k}x_{i,2}$$

the coefficients of the linear predictors have been set to:

$$\{\beta^*_0, \beta^*_1, \beta^*_2\}_k = \begin{cases} 0 & -0.3 & 0 & k = 1 \\ -0.5 & -0.9 & 0.4 & k = 2 \\ 0.5 & 0.4 & -0.3 & k = 3 \end{cases} \qquad \{\gamma^*_0, \gamma^*_1, \gamma^*_2\}_k = \begin{cases} 0 & 0 & -1 & k = 1 \\ 0.2 & 0 & 0.5 & k = 2 \\ 0.5 & 0 & 0.3 & k = 3 \end{cases}$$

For each parameter configuration, 200 samples have been considered. For estimation purpose we considered $k = 1, \dots, 5$ components and the best solution was chosen according to BIC. We report in Table 1 the values of the Rand index [20] between the original membership and the classification obtained using the CWR approach. As expected, with increasing noise, the quality of the classification deteriorates, decreasing from 0.98 for $\sigma^2 = 0.09$ to 0.80 for $\sigma^2 = 0.29$. As shown in the column *standard deviation* in Table 1, the variability of the Rand index distribution tends to increase as well. Overall, the coherence between the original classification and the one obtained using the proposed approach is quite satisfactory especially considering that our aim was to derive accurate estimates.

We have calculated the empirical coverage of the confidence intervals defined at the nominal level $1 - \alpha = 0.95$ for parameter estimates of the beta regression model, defined as the proportions of samples where the confidence intervals include the true parameter values in both scenarios (Tables 2 and 3). We do not observe a strong effect of the noise variance on the coverage, as the empirical proportions are all very close to the nominal confidence level even for large values of the noise variance.

Table 1 Average Rand index and corresponding standard deviation by values of σ^2

σ^2	Average Rand	Standard deviation
0.09	0.98	0.0047
0.11	0.97	0.0064
0.13	0.95	0.0076
0.15	0.94	0.0087
0.17	0.92	0.0100
0.19	0.90	0.0111
0.21	0.88	0.0118
0.23	0.86	0.0131
0.25	0.84	0.0134
0.27	0.82	0.0143
0.29	0.80	0.0133

Table 2 Empirical coverage of confidence intervals for model parameters: Scen. A

σ^2	Component 1	Component 2	Component 3
0.09	0.95	0.93	0.95
0.11	0.97	0.94	0.94
0.13	0.97	0.96	0.96
0.15	0.97	0.94	0.95
0.17	0.95	0.93	0.97
0.19	0.94	0.95	0.96
0.21	0.95	0.95	0.94
0.23	0.95	0.94	0.94
0.25	0.93	0.96	0.94
0.27	0.95	0.95	0.95
0.29	0.95	0.93	0.94

Table 3 Empirical coverage of confidence intervals for model parameters: Scen. B

ρ	b	Component 1	Component 2	Component 3
0.0	0.09	0.97	0.94	0.96
0.0	0.11	0.91	0.96	0.95
0.0	0.13	0.95	0.95	0.95
0.2	0.09	0.95	0.94	0.93
0.2	0.11	0.95	0.96	0.94
0.2	0.13	0.94	0.97	0.96
0.8	0.09	0.96	0.93	0.93
0.8	0.11	0.96	0.93	0.93
0.8	0.13	0.95	0.91	0.96

5 Concluding Remarks

We discuss cluster-weighted Beta regression to model the location and the precision parameter for a response with a conditional beta distribution and a multivariate Gaussian set of observed covariates. The proposal embeds the finite mixture of beta regressions as a particular case, see [10], when $g(\mathbf{x}_i \mid z_{ik} = 1) = g(\mathbf{x}_i)$, that is when the distribution of the observed covariates does not change across components. In fact, in that case, the likelihood for model parameters can be factorized, and it does not depend on the observed covariates distribution. While this is quite clear from a theoretical point of view, there can be some shortcomings when dealing with the implementation of the proposed method. Namely, the adopted objective function depends on the "scale" of the observed quantities and, when no clustering on the covariates is present, a single multivariate Gaussian component is well more parsimonius than a finite mixture of Gaussian components which is what would be implied by the finite mixture of the beta regression. Therefore the fact that the proposed method suggests $K = 1$ may well not be a signal that the beta regression model is homogeneous but, rather, that the Gaussian component is.

References

1. Bagnato, L., Punzo, A.: Finite mixtures of unimodal beta and gamma densities and the k-bumps algorithm. Comput. Stat. **28**, 1571–1597 (2013)
2. Buntaine, M.T.: Does the asian development bank respond to past environmental performance when allocating environmentally risky financing. World Dev. **39**, 336–350 (2011)
3. Carmichael, C.S.: Modelling student performance in a tertiary preparatory course. Master dissertation, University of Southern Queensland (2006)
4. Chen, S.X.: Beta kernel estimators for density functions. Comput. Stat. Data Anal. **31**, 131–145 (1999)
5. Cribari-Neto, F., Zeileis, A.: Beta regression in R. J. Stat. Softw. **34**(2), 1–24 (2010)
6. De Paola, M., Scoppa, V., Lombardo, R.: Can gender quotas break down negative stereotypes? Evidence from changes in electoral rules. J. Public Econ. **94**, 344–353 (2010)
7. Demsetz, H., Lehn, K.: The structure of corporate ownership: causes and consequences. J. Polit. Econ. **93**(6), 1155–1177 (1985)
8. Ferrari, S., Cribari-Neto, F.: Beta regression for modelling rates and proportions. J. Appl. Stat. **31**(7), 799–815 (2004)
9. Gershenfeld, N.: Nonlinear inference and cluster-wieghted modeling. Ann. N. Y. Acad. Sci. **808**(1), 18–24 (1997)
10. Grün, B., Kosmidis, I., Zeileis, A.: Extended beta regression in R: shaken, stirred, mixed, and partitioned. J. Stat. Softw. **48**(11), 1–25 (2012)
11. Hennig, C.: Identifiablity of models for clusterwise linear regression. J. Classif. **17**, 273–296 (2000)
12. Hunger, M., Doring, A., Holle, R.: Longitudinal beta regression models for analyzing health-related quality of life score overtime. BMC Med. Res. Methodol. **12**, 144 (2012)
13. Ingrassia, S., Punzo, A., Vittadini, G., Minotti, S.C.: The generalized linear mixed cluster-weighted model. J. Classif. **32**, 85–113 (2015)
14. Kieschnick, R., McCullough, B.D.: Regression analysis of variates observed on (0, 1): percentages, proportions and fractions. Stat. Model. **3**(3), 193–213 (2003)
15. Ospina, R., Ferrari, S.L.P.: Inflated beta distributions. Stat. Pap. **51**(1), 111 (2010)
16. Ospina, R., Ferrari, S.L.P.: A general class of zero-or-one inflated beta regression models. Comput. Stat. Data Anal. **56**(6), 1609–1623 (2012)
17. Papke, L., Wooldridge, J.: Econometric methods for fractional response variables with an application to 401(k) plan participation rates. J. Appl. Economet. **11**(6), 619–632 (1996)
18. Punzo, A., Ingrassia, S.: Clustering bivariate mixed-type data via the cluster-weighted model. Comput. Stat. **31**, 989–1013 (2016)

19. Punzo, A., Bagnato, L., Maruotti, A.: Compound unimodal distributions for insurance losses. Insur. Math. Econ. **81**, 95–107 (2018)
20. Rand, W.M.: Objective criteria for the evaluation of clustering methods. J. Am. Stat. Assoc. **66**(336), 846–850 (1971)
21. Rogers, J.A., Polhamus, D., Gillespie, W.R., Ito, K., Romero, K., Qiu, R., Stephenson, D., Gastonguay, M.R., Corrigan, B.: Combining patient-level and summary-level data for alzheimer's disease modeling and simulation: a beta regression meta-analysis. J. Pharmacokinet. Pharmacodyn. **39**, 479–498 (2012)
22. Swearingen, C.J., Tilley, B.C., Adams, R.J., Rumboldt, Z., Nicholas, J.S., Bandyopadhyay, D., Woolson, R.F.: Application of beta regression to analyze ischemic stroke volume in NINDS rt-PA clinical trials. Neuroepidemiology **37**(2), 73–82 (2011)

Detecting Wine Adulterations Employing Robust Mixture of Factor Analyzers

Andrea Cappozzo and Francesca Greselin

Abstract An authentic food is one that is what it claims to be. Nowadays, more and more attention is devoted to the food market: stakeholders, throughout the value chain, need to receive exact information about the specific product they are commercing with. To ascertain varietal genuineness and distinguish potentially doctored food, in this paper we propose to employ a robust mixture estimation method. Particularly, in a wine authenticity framework with unobserved heterogeneity, we jointly perform genuine wine classification and contamination detection. Our methodology models the data as arising from a mixture of Gaussian factors and depicts the observations with the lowest contributions to the overall likelihood as illegal samples. The advantage of using robust estimation on a real wine dataset is shown, in comparison with many other classification approaches. Moreover, the simulation results confirm the effectiveness of our approach in dealing with an adulterated dataset.

Keywords Mixtures of factor analyzers · Food authenticity · Model-based clustering · Wine adulteration · Robust estimation · Impartial trimming

1 Introduction and Motivation

The wine segment is identified as a luxury market category, with savvy as well as non-expert customers willing to spend a premium price for a product of a specific vintage and cultivar. Therefore, in the context of global markets, analytical methods for wine identification are needed in order to protect wine quality and prevent its illegal adulteration.

A. Cappozzo (✉) · F. Greselin
Department of Statistics and Quantitative Methods, University of Milano-Bicocca, Milano, Italy
e-mail: a.cappozzo@campus.unimib.it; francesca.greselin@unimib.it

F. Greselin et al. (eds.), *Statistical Learning of Complex Data*,
Studies in Classification, Data Analysis, and Knowledge Organization,
https://doi.org/10.1007/978-3-030-21140-0_2

In the present work we employ an approach based on robust estimation of mixtures of Gaussian Analyzers, for discriminating corrupted red wines samples from their authentic variety. In a modeling context, we assume a probability distribution function for the chemical and physical characteristics measured on the wines, considering a density in the form of a mixture, whenever the dataset presents more than a wine variety. As a consequence, the probability that a wine sample comes from one specific grape can be estimated from the model, performing classification through the Bayes rule. Robust estimation of the parameters in the model is adopted to recognize the corrupted data. Particularly, we expect that adulterated observations would be implausible under the robustly estimated model: the illegal subsample is revealed by selecting observations with the lowest contributions to the overall likelihood using impartial trimming, without imposing any assumption on their underlying density.

The rest of the paper is organized as follows: in Sect. 2 the notation is introduced and the main concepts about Gaussian Mixtures of Factor Analyzers (MFA), trimmed MFA likelihood, and the Alternating Expectation-Conditional Maximization (AECM) algorithm are summarized. Section 3 presents the *wine* dataset [7] and classification results obtained performing a robust estimation of Gaussian mixtures of factor analyzers. Section 4 reports a simulation study carried out employing parameters estimated from the model in Sect. 3, in a specific framework of contaminated dataset.

The original contribution of the present paper is given in the benchmark study on unsupervised methods, the adaptation of the robust Bayesian Information Criterion (BIC) introduced in [3] to MFA, and a first application of robust MFA in a somehow realistic adulteration scenario.

An application on real data and some simulation results confirm the effectiveness of our approach in dealing with an adulterated dataset when compared to analogous methods, such as partition around medoids and non-robust mixtures of Gaussian and mixtures of patterned Gaussian factors.

2 Mixtures of Gaussian Factors Analyzers

In this section we briefly recall the definition and some features of the mixture of Gaussian Factor Analyzers (MFA) and its parameter estimation procedure. MFA is a powerful tool for modeling unobserved heterogeneity in a population, as it concurrently performs clustering and local dimensionality reduction, within each cluster. Let $\mathbf{X}_1, \ldots, \mathbf{X}_n$ be a random sample of size n on a p-dimensional random vector. An MFA assumes that each observation $\mathbf{X}_i$ is given by

$$\mathbf{X}_i = \boldsymbol{\mu}_g + \boldsymbol{\Lambda}_g \mathbf{U}_{ig} + \mathbf{e}_{ig} \tag{1}$$

with probability π_g for $g = 1, \ldots, G$. The total number of components in the mixture is denoted by G, $\boldsymbol{\mu}_g$ are $p \times 1$ mean vectors, $\boldsymbol{\Lambda}_g$ are the $p \times d$ matrices of *factor loadings*, $\mathbf{U}_{ig} \overset{iid}{\sim} \mathcal{N}(\mathbf{0}, \mathbf{I}_d)$ are the *factors*, $\mathbf{e}_{ig} \overset{iid}{\sim} \mathcal{N}(\mathbf{0}, \boldsymbol{\Psi}_g)$ are the *errors*, and $\boldsymbol{\Psi}_g$ are $p \times p$ diagonal matrices. Note that $d < p$, that is the p observable features are supposed to be jointly explained by a smaller number of d unobservable factors. Further, $\mathbf{U}_{ig}$ and $\mathbf{e}_{ig}$ are independent, for $i = 1, \ldots, n$ and $g = 1, \ldots G$. Unconditionally, therefore, $\mathbf{X}_i$ has a density in the form of a G-components multivariate normal mixture:

$$f_{\mathbf{X}_i}(\mathbf{x}_i; \boldsymbol{\theta}) = \sum_{g=1}^{G} \pi_g \phi_p(\mathbf{x}_i; \boldsymbol{\mu}_g, \boldsymbol{\Sigma}_g) \tag{2}$$

where $\phi_p(\cdot; \boldsymbol{\mu}_g, \boldsymbol{\Sigma}_g)$ denotes the p-multivariate normal density, whose covariance matrix $\boldsymbol{\Sigma}_g$ has the following decomposition $\boldsymbol{\Sigma}_g = \boldsymbol{\Lambda}_g \boldsymbol{\Lambda}_g' + \boldsymbol{\Psi}_g$.

When estimating MFA through the usual Maximum Likelihood approach, two issues arise. Firstly, departure from normality in the data may cause biased or misleading inference. Some initial attempts in the literature to overcome this issue propose to consider mixtures of t-factor analyzers [15], but the breakdown properties of the estimators are not improved [10]. The second concern is related to the unboundedness of the log-likelihood function [4], which leads to estimation issues, like the appearance of non-interesting *spurious maximizers* and degenerate solutions. To cope with this second issue, Common/Isotropic noise matrices/patterned covariances [1] and a mild constrained estimation [9] have been considered. The methodology considered here employs model estimation, complemented with *trimming* and *constrained estimation*, to provide robustness, to exclude singularities, and to reduce spurious solutions, along the lines of [8]. Therefore, with this approach, we overcome both previously mentioned issues.

A mixture of Gaussian factor components is fitted to a given dataset $\mathbf{x}_1, \mathbf{x}_2, \ldots, \mathbf{x}_n$ in $\mathbb{R}^p$ by maximizing a *trimmed mixture log-likelihood* [18],

$$\mathcal{L}_{trim} = \sum_{i=1}^{n} \zeta(\mathbf{x}_i) \log \left[\sum_{g=1}^{G} \phi_p(\mathbf{x}_i; \mu_g, \boldsymbol{\Lambda}_g, \boldsymbol{\Psi}_g) \pi_g \right] \tag{3}$$

where $\zeta(\cdot)$ is a 0–1 trimming indicator function that tells us whether observation $\mathbf{x}_i$ is trimmed off or not. If $\zeta(\mathbf{x}_i)$=0 $\mathbf{x}_i$ is trimmed off, otherwise $\zeta(\mathbf{x}_i)$=1. A fixed fraction α of observations, the *trimming level*, is unassigned by setting $\sum_{i=1}^{n} \zeta(\mathbf{x}_i) = \lceil n(1-\alpha) \rceil$, where the less plausible observations under the currently estimated model are tentatively trimmed out at each step of the iterations that lead to the final estimate. In the specific application to wine authenticity analysis described in Sect. 3, they are supposed to be originated by wine adulteration.

Then, a constrained maximization of (3) is adopted, by imposing $\psi_{g,ll} \leq c\,\psi_{h,mm}$ for $1 \leq l \neq m \leq p$ and $1 \leq g \neq h \leq G$, where $\{\psi_{g,ll}\}_{l=1,\ldots,p}$ are the diagonal element of the noise matrices $\boldsymbol{\Psi}_g$, and $1 \leq c < +\infty$, to avoid the $|\Sigma_g| \to 0$ case. This constraint can be seen as an adaptation to MFA of those introduced in [11]. The Maximum Likelihood estimator of $\boldsymbol{\Psi}_g$ under the given constraints leads to a well-defined maximization problem.

The Alternating Expectation-Conditional Maximization—an extension of the Expectation-Maximization algorithm—is considered, in view of the factor structure of the model. The M-step is replaced by some computationally simpler conditional maximization (CM) steps, along with different specifications of missing data. The idea is to partition the vector of parameters $\boldsymbol{\theta} = (\boldsymbol{\theta}_1', \boldsymbol{\theta}_2')'$, in such a way that $\mathcal{L}_{trim}$ is easy to be maximized for $\boldsymbol{\theta}_1$ given $\boldsymbol{\theta}_2$ and vice versa. Therefore, two cycles are performed at each algorithm iteration:

1st*cycle* : we set $\boldsymbol{\theta}_1 = \{\pi_g, \boldsymbol{\mu}_g, g = 1, \ldots, G\}$; here, the missing data are the unobserved group labels $\mathbf{Z} = (\mathbf{z}_1', \ldots, \mathbf{z}_n')$. After applying a step of Trimming, by assigning to the observations with lowest likelihood a null value of the "posterior probabilities", we get one E-step, and one CM-step for obtaining parameters in $\boldsymbol{\theta}_1$.

2nd*cycle* : we set $\boldsymbol{\theta}_2 = \{\boldsymbol{\Lambda}_g, \boldsymbol{\Psi}_g, g = 1, \ldots, G\}$, here the missing data are the group labels $\mathbf{Z}$ and the unobserved latent factors $\mathbf{U}_{11}, \ldots, \mathbf{U}_{nG}$. We perform a Trimming step, then a E-step, and a constrained CM-step, i.e., a conditional exact constrained maximization of $\boldsymbol{\Lambda}_g, \boldsymbol{\Psi}_g$.

A detailed description of the algorithm is given in [8].

3 Wine Recognition Data

The wine recognition dataset, firstly analysed in [7], reports results of a chemical and physical analysis for three different wine types, grown in the same region in Italy. Originally, 28 attributes were recorded for 178 wine samples derived from three different cultivars: Barolo, Grignolino, and Barbera. A reduced version of the original dataset with only thirteen variables is publicly available in the University of California, Irvine Machine Learning data repository, commonly used in testing the performance of newly introduced supervised and unsupervised classifiers. Particularly, in the unsupervised classification literature the wine recognition data has been considered to assess cluster analysis in information-theoretic terms via minimisation of the partition entropy [19], to prove the modelling capabilities of a generalized Dirichlet mixture [2], to evaluate the efficacy of employing distances based on non-Euclidean norms [5] and of Random Forest dissimilarity [20]. More recently, also parsimonious Gaussian mixture models have been applied to the Italian wines dataset [16].

Here our purpose is twofold: we want to explore the classification performance of a robust estimation based on mixtures of Gaussian Factors Analyzers, and we aim at obtaining realistic parameters for the subsequent simulation study. The dataset, available in the *pgmm* R package [17], contains 27 of the 28 original variables, since the sulphur measurements were not available. Initially, to perform model selection and detect the most suitable values of factors d and groups G, an adaptation to the

Table 1 *RobustBIC* [3] for different choices of the number of factors d and the number of groups G for the robust MFA model on wine data, trimming level $\alpha = 0.05$ and $c = 20$. The smallest value is obtained with $d = 4$ and $G = 2$

	G		
d	1	**2**	3
1	9082.58	8282.92	8223.46
2	8560.62	8107.62	8112.90
3	8352.26	8042.02	8199.38
4	8160.77	**7969.64**	8315.23
5	8102.77	8044.03	8456.00
6	8097.06	8165.67	8735.63

Table 2 Classification table for the robust MFA with number of factors $d = 4$, number of groups $G = 3$, trimming level $\alpha = 0.05$ and $c = 20$ on the wine data

	1	2	3
Barolo	59	0	0
Grignolino	0	71	0
Barbera	0	0	48

Trimmed observations are classified a posteriori according to the Bayes rule

MFA framework of the robust Bayesian Information Criterion, firstly introduced in [3], has been considered. That is, $BIC = -2\mathcal{L}_{trim}(x;\hat{\theta}) + v^c \log n^*$ where $v^c = (G - 1 + Gp + G(pd - d(d-1)/2) + (Gp-1)(1-1/c) + 1)$ denotes the number of free parameters in the model (depending on the value of the constraint c) and $n^* = \lceil n(1-\alpha) \rceil$ the number of non-trimmed observations. Robust BIC for different choices of the number of factors d and the number of groups G are reported in Table 1, considering a trimming level $\alpha = 0.05$ and $c = 20$. The value of the robust BIC is minimized for $d = 4$ and $G = 2$, suggesting a mixture with just two components. Careful investigation on this result highlighted that robust MFA methodology tended to cluster together Barolo and Grignolino samples as arising from the same mixture component, while clearly separating Barbera observations. It is worth recalling [7] that the wines in this study were collected over the time period of 1970–1979, and the Barbera wines are predominantly from a later period than the Barolo or Grignolino wines. Therefore, considering the nature of the phenomena under study and the risks related to rigidly selecting the number of components in a mixture model only on the basis of the results provided by an information criteria, such as BIC [13], we decided to employ a robust MFA with $d = 4$, $G = 3$, and $\alpha = 0.05$, leading to the classification matrix reported in Table 2. Employing a robust MFA rather than a Gaussian mixture leads to a 60% reduction in the number of parameters to be estimated (470 against 1217). Notice, in addition, that after robust estimation, also the trimmed observations can be a posteriori classified according to the Bayes rule, i.e., assigning each of them to the component g having greater value of $D_g(\mathbf{x}, \theta) = \phi_p(\mathbf{x}; \mu_g, \boldsymbol{\Lambda}_g \boldsymbol{\Lambda}_g^{'} + \boldsymbol{\Psi}_g)\pi_g$.

Results in Table 2 show that the robust MFA algorithm led to a perfect clusterization of the samples according to their true wine type.

For completeness, the robust MFA algorithm was also applied to the more common thirteen variable subset of the wine data and comparison with the existing literature is reported in Table 3. The clustering performance with respect to the true

Table 3 Comparison of performance metrics for different methodologies on the thirteen variable subset of the wine data

Methodology	Performance metric	
	Class recovery accuracy	Adjusted Rand index
Partition entropy [19]	0.977	–
Mixture of generalized Dirichlet [2]	0.978	–
Neural gas [5]	0.954	–
Random Forest predictors [20]	–	0.93
Parsimonious Gaussian mixture [16]	0.927	0.79
Robust MFA [8]	0.994	0.98

Reported metrics come from the original articles

wine labels reports an *Adjusted Rand Index* equal to 0.98 with just one Grignolino sample wrongly assigned to the cluster identifying Barolo wines. Again then, the robust MFA methodology outperforms the results currently present in the literature for unsupervised learning on this specific dataset.

4 Simulation Study

The purpose of this simulation study is to show the effectiveness of estimating a robust MFA on a set of observations drawn from two luxury wines, Barolo and Grignolino, and identifying units presenting an adulteration. Considering the parameters estimated obtained in Sect. 3, the artificial dataset is generated simulating 100 observations each, from Barolo and Grignolino components. Afterwards, the "contamination" is created decreasing by 15% the values of Fixed Acidity, Tartaric Acid, Malic Acid, Uronic Acids, Potassium, and Magnesium for 5 Barolo and for 5 Grignolino observations. This procedure resembles the illegal practice of adding water to wine [12]. The problem of distinguishing adulterated observations from the real mixture components is addressed, together with the algorithm performance in correctly classifying the authentic units.

We estimate a robust MFA with $G = 2$, $p = 27$, $d = 4$ and trimming level $\alpha = 0.05$. We compare our results with other popular methods: Partition around medoids, Gaussian mixtures estimated via Mclust, and Mixtures of patterned Gaussian factors estimated by $pgmm$. To perform each of the $B = 1000$ simulations, algorithms have been initialized following the indications of their respective authors: say 10 random starts at each run of $AECM$, default setting for the "build phase" of pam as in [14], applying model-based hierarchical clustering as per default setting in [6] for $Mclust$ and 10 random starts at each run as suggested in [16] for $pgmm$.

Table 4 Average misclassification errors and ARI (percent average values on 1000 runs)

	AECM	*pam*	*Mclust*	*pgmm*
Misclassification error	0.0309	0.2935	0.2073	0.2314
Adjusted Rand Index	0.9362	0.5466	0.7184	0.6959

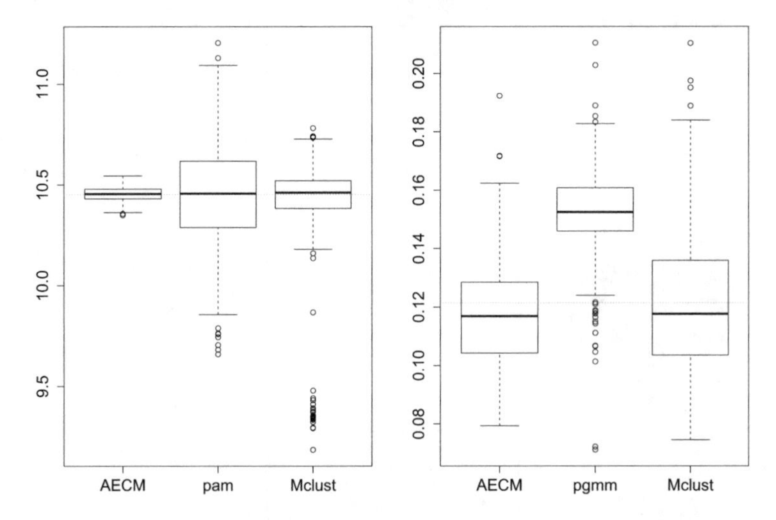

Fig. 1 Boxplots of the simulated distributions of $\hat{\boldsymbol{\mu}}_1[1]$, estimator for $\boldsymbol{\mu}_1[1] = 10.45$ (left panel); $\hat{\Sigma}_1[1, 1]$, estimator for $\Sigma_1[1, 1] = 0.1214$ (right panel)

Table 4 reports the average misclassification error and Adjusted Rand Index: the AECM algorithm reports a superb classification rate, with smaller variability of the simulated distributions for the estimated quantities, as shown in Fig. 1.

For a fair comparison of the performance of the algorithms, we consider 3 clusters for *pam*, *Mclust*, and *pgmm*; whereas we consider only 2 clusters for AECM, because in this approach the adulterated group should ideally be captured by the trimmed units. A value of $c = 20$ allows to discard singularities and to reduce spurious solutions [8]. The effects of the trimming procedure are shown in Fig. 2, where the different colours and shapes represent the obtained classification. Table 5 reports the average bias and MSE for the mixture parameters (computed element-wise for every component). While an R package is under construction, R scripts containing the employed routines are available from the authors upon request.

The present simulations show initial promising results in adopting robust MFA as a tool for identifying wine adulteration. Future research regards a novel approach for semi-supervised robust clustering, allowing for impartial trimming on both labelled and unlabelled data partitions. The aim is to jointly address methodological issues in robust statistics and clustering, as well as providing consistent statistical tools required in the increasingly demanding food authenticity domain.

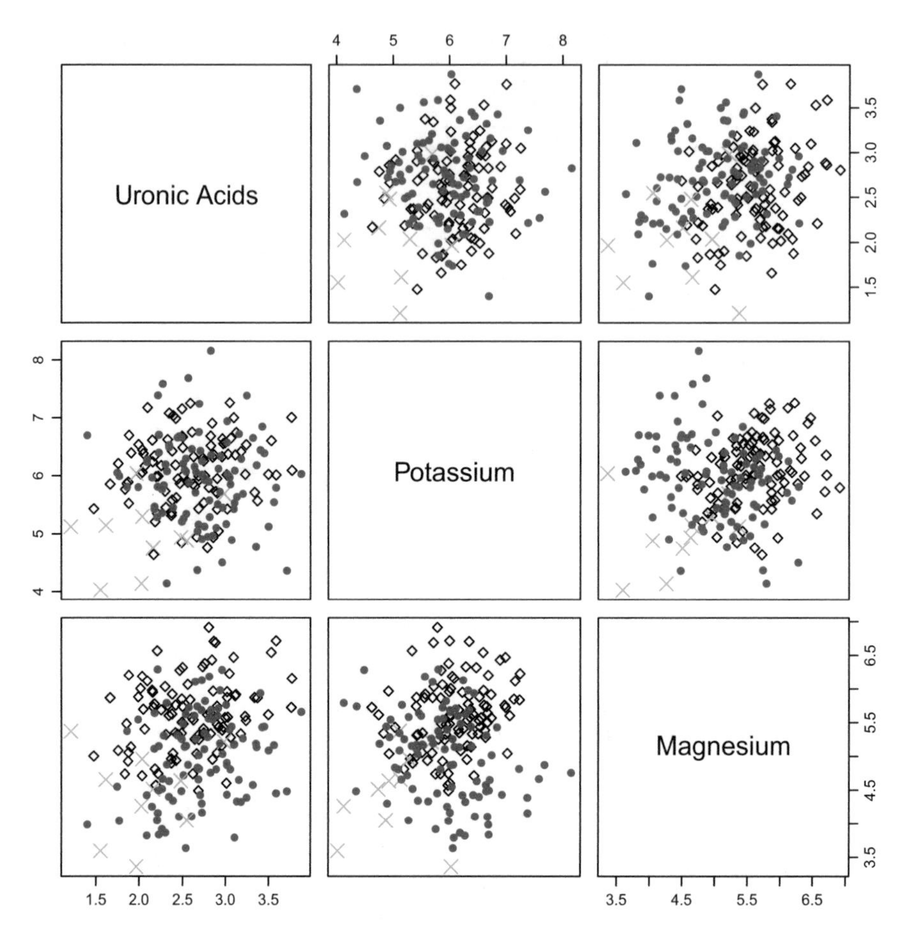

Fig. 2 Clustering of the simulated data with fitted trimmed and constrained MFA. Trimmed observations are denoted by "×"

Table 5 Bias and MSE (in parentheses) of the parameter estimators $\hat{\boldsymbol{\mu}}_g$ and $\hat{\Sigma}_g$

	$AECM$	$Mclust$	pam		$AECM$	$Mclust$	$pgmm$
$\boldsymbol{\mu}_1$	−0.0019	−0.0194	0.0069	Σ_1	0.0001	−0.001	0.0257
	(0.0029)	(0.0421)	(0.1022)		(0.0004)	(0.0022)	(0.0079)
$\boldsymbol{\mu}_2$	−0.0011	0.1522	−0.0025	Σ_2	−0.0156	−0.0164	0.0113
	(0.0042)	(0.2376)	(0.1380)		(0.0043)	(0.0043)	(0.0077)

References

1. Baek, J., McLachlan, G.J., Flack, L.K.: Mixtures of factor analyzers with common factor loadings: applications to the clustering and visualization of high-dimensional data. IEEE Trans. Pattern Anal. Mach. Intell. **32**(7), 1298–1309 (2010)
2. Bouguila, N., Ziou, D.: A powerful finite mixture model based on the generalized Dirichlet distribution: unsupervised learning and applications. In: Proceedings of the 17th International Conference on Pattern Recognition, ICPR 2004, vol. 1, pp. 280–283. IEEE, Piscataway (2004)
3. Cerioli, A., García-Escudero, L.A., Mayo-Iscar, A., Riani, M.: Finding the number of normal groups in model-based clustering via constrained likelihoods. J. Comput. Graph. Stat. **27**(2), 404–416 (2018)
4. Day, N.E.: Estimating the components of a mixture of normal distributions. Biometrika **56**(3), 463–474 (1969)
5. Doherty, K.A.J., Adams, R.G., Davey, N.: Unsupervised learning with normalised data and non-Euclidean norms. Appl. Soft Comput. **7**(1), 203–210 (2007)
6. Fop, M., Murphy, T.B., Raftery, A.E.: mclust 5: clustering, classification and density estimation using gaussian finite mixture models. R J. **XX**(August), 1–29 (2016)
7. Forina, M., Armanino, C., Castino, M., Ubigli, M.: Multivariate data analysis as a discriminating method of the origin of wines. Vitis **25**(3), 189–201 (1986)
8. García-Escudero, L.A., Gordaliza, A., Greselin, F., Ingrassia, S., Mayo-Iscar, A.: The joint role of trimming and constraints in robust estimation for mixtures of Gaussian factor analyzers. Comput. Stat. Data Anal. **99**, 131–147 (2016)
9. Greselin, F., Ingrassia, S.: Maximum likelihood estimation in constrained parameter spaces for mixtures of factor analyzers. Stat. Comput. **25**(2), 215–226 (2015)
10. Hennig, C.: Breakdown points for maximum likelihood estimators of location-scale mixtures. Ann. Stat. **32**(4), 1313–1340 (2004)
11. Ingrassia, S.: A likelihood-based constrained algorithm for multivariate normal mixture models. Stat. Methods Appl. **13**(2), 151–166 (2004)
12. Jackson, R.S.: Wine Science: Principles and Application. Academic press, Elsevier (2008)
13. Lee, S.X., McLachlan, G.J.: Finite mixtures of canonical fundamental skew t-distributions: the unification of the restricted and unrestricted skew t-mixture models. Stat. Comput. **26**(3), 573–589 (2016)
14. Maechler, M., Rousseeuw, P., Struyf, A., Hubert, M., Hornik, K.: cluster: Cluster analysis basics and extensions, R package version 2.1.0 – For new features, see the 'Changelog' file (in the package source) (2019)
15. McLachlan, G.J., Bean, R.W., Ben-Tovim Jones, L.: Extension of the mixture of factor analyzers model to incorporate the multivariate t-distribution. Comput. Stat. Data Anal. **51**(11), 5327–5338 (2007)
16. McNicholas, P.D., Murphy, T.B.: Parsimonious Gaussian mixture models. Stat. Comput. **18**(3), 285–296 (2008)
17. McNicholas, P.D., ElSherbiny, A., McDaid, A.F., Murphy, T.B.: pgmm: Parsimonious Gaussian mixture models, R package version 1.2.3. https://CRAN.R-project.org/package=pgmm (2018)
18. Neykov, N., Filzmoser, P., Dimova, R., Neytchev, P.: Robust fitting of mixtures using the trimmed likelihood estimator. Comput. Stat. Data Anal. **52**(1), 299–308 (2007)
19. Roberts, S.J., Everson, R., Rezek, I.: Maximum certainty data partitioning. Pattern Recognit. **33**(5), 833–839 (2000)
20. Shi, T., Horvath, S.: Unsupervised learning with random forest predictors. J. Comput. Graph. Stat. **15**(1), 118–138 (2006)

Simultaneous Supervised and Unsupervised Classification Modeling for Assessing Cluster Analysis and Improving Results Interpretability

Mario Fordellone and Maurizio Vichi

Abstract In the unsupervised classification field, the unknown number of clusters and the lack of assessment and interpretability of the final partition by means of inferential tools denote important limitations that could negatively influence the reliability of the final results. In this work, we propose to combine unsupervised classification with supervised methods in order to enhance the assessment and interpretation of the obtained partition. In particular, the approach consists in combining of the clustering method k-means (KM) with logistic regression (LR) modeling to have an algorithm that allows an evaluation of the partition identified through KM, to assess the correct number of clusters, and to verify the selection of the most important variables. An application on real data is presented to better clarify the utility of the proposed approach.

Keywords Supervised classification · Unsupervised classification · Assessing clustering

1 Introduction

In unsupervised classification techniques, clusters of homogeneous objects are detected by means of a set of features measured (observed) on a set of objects without knowing the membership of objects to clusters. In these applications, the aim is to discover the heterogeneous structure of the data. In unsupervised classification models, the principal approaches of cluster analysis [6] are: connectivity-based clustering better known as hierarchical clustering, centroid-based clustering, distribution-based clustering, density-based clustering, and many other parametric and non-parametric techniques [7].

M. Fordellone (✉) · M. Vichi
Department of Statistical Sciences, Sapienza University of Rome, Rome, Italy
e-mail: mario.fordellone@uniroma1.it; maurizio.vichi@uniroma1.it

F. Greselin et al. (eds.), *Statistical Learning of Complex Data*,
Studies in Classification, Data Analysis, and Knowledge Organization,
https://doi.org/10.1007/978-3-030-21140-0_3

Conversely, supervised classification is based on the idea of forecasting the membership of new objects (output) based on a set of features (inputs) measured on a training set of objects for which the membership to clusters is known. Therefore, in these applications, the aim is to generalize a function or mapping from inputs to outputs which can then be used speculatively to generate an output for previously unseen inputs [4, 8]. Usually, a subsample (training), which is representative of specific groups, is selected and then this model is used as reference for the classification of new (unobserved) other objects. Training sets are selected based on the knowledge of the user. In supervised classification models we have artificial neural networks, naive Bayes classifiers, nearest neighbor algorithm naive, decision trees, logistic regression, generalized linear models, and many other parametric and non-parametric techniques.

In unsupervised classification, we have important issues that could drastically influence results: (1) an unknown number of clusters, (2) an absence of variable selection that most contribute to clustering, and (3) a final assessment of clusters [3]. In other words, all the decisions taken to address the study can lead to different results and each single decision becomes crucial for the aim of our study and needs to be tested.

In this work, we propose an algorithm based on the use of supervised classification modeling. In particular, our approach consists in the simultaneous application of k-means (KM) [10] and logistic regression (LR) [1] modeling. We will prove that, by using LR, we have effective inferential tools for choosing the number of clusters, selecting the most important variables for the clustering, and assessing the quality of clusters.

The paper is structured as follows: in Sect. 2 we present our proposal for the simultaneous application of unsupervised and supervised classification modeling, in Sect. 3 we show an application on real data, and finally, in Sect. 4 we try to give some suggestions and concluding remarks on the work.

2 Proposal

In unsupervised classification modeling, we are not interested in prediction because we do not have an associated response variable y like in supervised classification modeling. Therefore, this paper proposes to simultaneously apply unsupervised (i.e., KM) and supervised classification (i.e., LR) approaches, where the latter aims to evaluate and to improve the former with the additional data structure information. We will call this approach k-means-logistic regression (KM-LR). In particular, KM-LR is composed of the following principal steps:
Given the $n \times J$ data matrix $\mathbf{X}$, for $K = 2, \ldots, Kmax$, where $Kmax$ is the maximum number of clusters the researcher thinks the data might have, the algorithm works as follows:

Algorithm 1 KM-LR algorithm

1: **for** $k = 2$ to $Kmax$ **do**
2: **K-means step**
3: Randomly initialize the membership matrix $\mathbf{U}$;
4: Compute the centroids matrix by $\bar{\mathbf{X}} = (\mathbf{U}^T\mathbf{U})^{-1}\mathbf{U}^T\mathbf{X}$;
5: Minimize the objective function $\|\mathbf{X} - \mathbf{U}\bar{\mathbf{X}}\|^2$ with respect to the membership matrix $\mathbf{U}$;
6: Update the centroids matrix $\bar{\mathbf{X}}_\mathbf{n} = (\mathbf{U_n}^T\mathbf{U_n})^{-1}\mathbf{U_n}^T\mathbf{X}$ given the new assignment matrix $\mathbf{U_n}$;
7: **if** $\|\mathbf{X} - \mathbf{U_n}\bar{\mathbf{X}}_\mathbf{n}\|^2 > \omega$;
$\bar{\mathbf{X}} = \bar{\mathbf{X}}_\mathbf{n}$, $\mathbf{U} = \mathbf{U_n}$, repeat steps 5–6;
else
exit loop; obtain the g_k categorical cluster vector;
8: **end if**
9: **Multinomial logistic regression step**
10: LR is estimated on g_k, with explanatory variables $\mathbf{X}$, for estimating the probabilities for its $k-1$ response categories $\pi_k(\mathbf{x})$, and to estimate the probabilities for its *baseline* category $\pi_0(\mathbf{x})$;
11: **if** some LR estimated coefficient is not 5% statistically significant;
remove the corresponding variables from the matrix $\mathbf{X}$;
repeat steps 2–10;
12: **end if**
13: **end for**

At the end, we obtain $Kmax - 1$ identified partitions (with a different number of clusters k), together with a reduced set of statistically significant variables and a set of inferential tools to assess the quality of the partition. The best partition (with the optimal number of clusters k) is identified in correspondence of the largest increase of a χ^2-test computed on the partitions obtained by KM and LR. In this way, through the analysis of the LR results (e.g., explained variance, parameters significance, residual variance), we have an evaluation of the partition obtained by KM. In fact, a good performance of the LR model on the response variable derived by the KM outcome means that the variables included in the model provide a good explanation for the group structure in the data. Moreover, through the LR coefficients analysis, we can see which variables contribute the most to identifying the group structure and to what extent they do so (by analyzing statistical significance, value estimates, and signs of coefficients).

Note that the algorithm monotonically decreases the loss function or at least does not increase it. However, it does not guarantee to stop at the global minimum of the loss function. For this reason, it is recommended to use a large number of randomly started runs to find the best solution. The predictive accuracy of the methodology can be assessed by cross-validation to give an insight into how the model will generalize to an independent data set. In a following paper we will include a cross-validation procedure and a simulation study to assess the predictive accuracy and evaluate the performances of the algorithm.

In the next section, an application on real data is presented.

3 Application on Real Data

In this section a real data application of KM-LR is presented. The data set is named *Wine Data* [5]. It is the result of a chemical analysis of wines grown in an Italian region, derived from three different cultivars.

The 13 constituents were measured on 178 types of wine from the three cultivars: 59, 71, and 48 instances are in class one, two, and three, respectively. The 13 chemical continuous attributes of the wine data set are: 1. *Alcohol* (Alc), 2. *Malic acid* (Mal), 3. *Ash* (Ash), 4. *Alkalinity of ash* (AAsh), 5. *Magnesium* (Mg), 6. *Total phenols* (Phe), 7. *Flavonoids* (Fla), 8. *Non-Flavonoids phenols* (NPhe), 9. *Proanthocyanidins* (ProA), 10. *Color intensity* (Col), 11. *Hue* (Hue), 12. *OD280-OD315 of diluted wines* (ROD), and 13. *Proline* (Pro).

In the analysis, we have tried to select the optimal number of clusters without considering the a priori information that $K = 3$, and using the KM-LR algorithm, i.e., through the maximization of the increase of the χ^2-test computed on the partitions obtained by KM and LR. For comparison purposes, two other approaches have been used. The procedure has been randomly repeated 50 times from 2 to 10 clusters using a single random start. In Table 1, the results obtained by KM-LR (first column), the sequential application of KM followed by the *Gap-method* proposed by Tibshirani [12] (second column), and the sequential application of KM followed by Calinski and Harabasz's [2] criterion (third column) have been reported.
The best performance has been obtained by the KM-LR approach, where the optimal number of clusters has been captured 36 times out of 50 (72%) runs. In contrast, the KM-*Gap-method* obtained the worst performance, since the optimal number of clusters was captured only 5 times (10%). Thus, the KM-LR approach seems to reduce the effect of the local minima problem of the KM algorithm, and this is even more relevant in the case no modification of the KM partition as proposed by the KM –> *Gap-method* and KM –> *Calinski–Harabasz* method.

Table 1 Optimal K selection from 2 to 10 clusters on the 50 random repeat using a single random start

	KM-LR		KM –> Gap-method		KM –> Calinski–Harabasz	
K	Count	Percent	Count	Percent	Count	Percent
2	0	0.00	0	0.00	0	0.00
3	36	72.00	5	10.00	22	44.00
4	10	20.00	0	0.00	5	10.00
5	2	4.00	0	0.00	3	6.00
6	2	4.00	0	0.00	3	6.00
7	0	0.00	2	4.00	0	0.00
8	0	0.00	1	2.00	0	0.00
9	0	0.00	15	30.00	6	12.00
10	0	0.00	27	54.00	11	22.00
Total	50	100.00	50	100.00	50	100.00

Table 2 Estimation results obtained by logistic regression applied to the KM partition including only predictors with a 5% significant coefficient

	Estimate	SE	t-stat	p-value
Const.	2.0169	0.0296	68.2200	2.66E−122
Alc	−0.2306	0.0465	−4.9579	1.76E−06
Mal	−0.0865	0.0382	−2.2674	2.47E−02
Mg	−0.1264	0.0353	−3.5808	4.51E−04
Fla	−0.2012	0.0786	−2.5597	1.14E−02
Col	−0.0806	0.0516	−1.5634	1.20E−02
Hue	0.0970	0.0474	2.0492	4.20E−02
Pro	−0.3627	0.0498	−7.2806	1.31E−11

178 observations, 164 error degrees of freedom; Dispersion: 0.138, AICc = 160.34, BIC = 185.95; R-squared-adj. = 0.8135; F-statistic: 93.70, p-value = 5.19E−55

In Table 2 we show the estimation results of LR applied to the group labels identified through the KM model as a response variable and include only variables with significant coefficients as predictors.

From Table 2 we can note that the model shows good performance and about 80% of the total deviance is explained (i.e., $R^2_{adj} = 0.81$). The variables *Ash*, *Alkalinity of Ash*, *Total phenols*, *Non-Flavonoids phenols*, *Proanthocyanidins*, and the *OD280-OD315 of diluted wines* have been excluded because these were not statistically significant at the 5% level. In Fig. 1 the partitions identified by KM-LR (highlighted with different symbols) on the 7 included variables have been represented.

The partition seems well represented on most pairs of variables, because it is represented by the statistically most significant variables. Moreover, the partition found by the KM-LR approach better identifies the real data partition identified by the three different cultivars.

Table 3 shows (1) the confusion matrix between the real data partition and the KM partition (i.e., KM applied to the complete data) and (2) the confusion matrix between the real data partition and the KM-LR partition.

The misclassification rate and the adjusted Rand index (ARI) [11] applied on the left table (i.e., the real partition versus the KM partition) are equal to 0.3708 and 0.2977, respectively; these same indices applied to the right table (i.e., the real partition versus the KM-LR partition) are equal to 0.1818 and 0.5465, respectively. We recall that ARI has a value between 0 and 1, with 0 indicating that the two data clusterings do not agree on any pair of points and 1 indicating that the data clusterings are identical.

Moreover, by applying LR to the real data partition we obtain the following confusion matrix between the real partition and the one fitted by LR (Table 4).

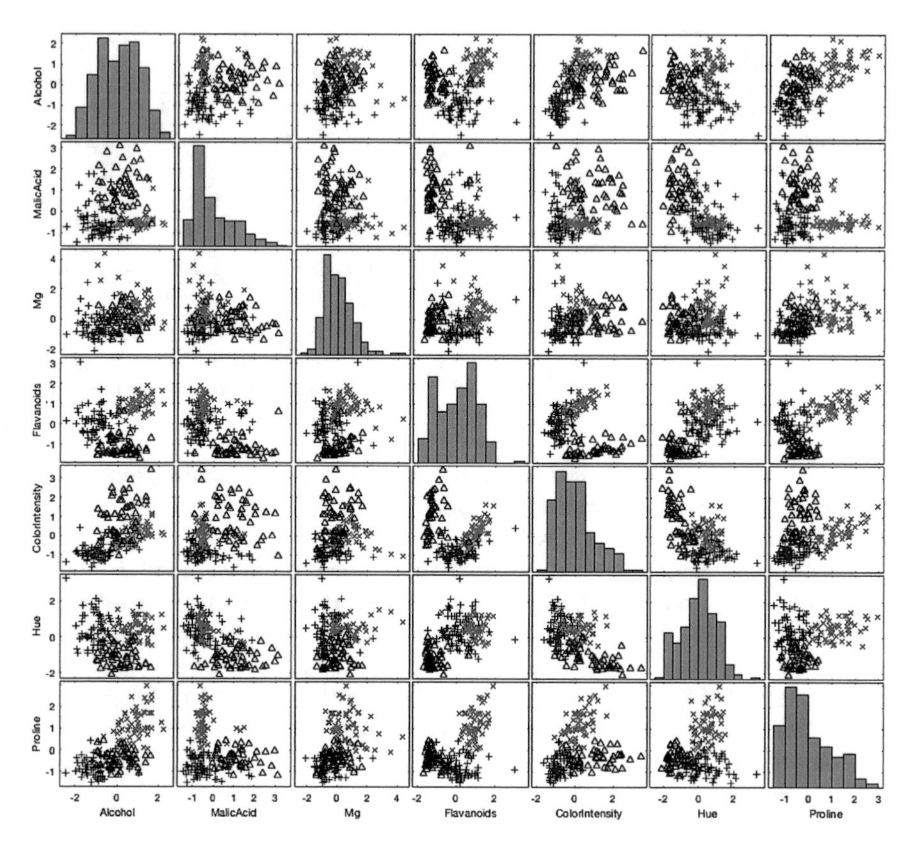

Fig. 1 The three clusters identified by KM-LR represented on the variables included in the model

Table 3 Confusion matrix between: (1) real data partition and KM partition; (2) real data partition and KM-LR partition

Real	K-means			Total	Real	K-means - LR			Total
	C_1	C_2	C_3			C_1	C_2	C_3	
C_1	32	5	22	59	C_1	51	3	5	59
C_2	9	61	1	71	C_2	3	66	2	71
C_3	2	27	19	48	C_3	0	12	36	48
Total	43	93	42	178	Total	54	81	43	178

Table 4 Confusion matrix between real data partition and LR partition

Real	Logistic regression			Total
	C_1	C_2	C_3	
C_1	15	44	0	59
C_2	6	62	3	71
C_3	2	38	8	48
Total	23	144	11	178

The performance of KM-LR is also better. In fact, the misclassification rate and ARI applied to Table 4 are equal to 0.5225 and 0.0247, respectively. In Table 5, the performances obtained by both LR applied to the real partition and KM-LR are shown. We note that the diagnostic indices obtained by KM-LR are much than those obtained by the LR application on the real data partition. Furthermore, in the application of LR on the real data partition, only the variable *Color intensity* has obtained a statistically significant coefficient and then, only this variable has been included in the model.

Finally, to obtain a quality measure of the clusters, a MANOVA model [9] on the real data partition and on that obtained by the KM and KM-LR models has been applied (Table 6).

The null hypothesis is rejected in each of the three cases, i.e., the means of each group are not the same j-dimensional multivariate vector, and any difference observed in the sample is not due to random chance. However, we can note that the most significant value of λ is derived in the KM-LR partition. In Fig. 2, the distributions of the three KM-LR clusters on the reduced set of variables are shown.

Table 5 Comparison between LR and KM-LR

	LR	KM-LR
F-Statistic	14.5000	93.7000
p-value	0.0002	5.19E−55
R-squared-adj.	0.0710	0.8135
AICc	403.3673	160.3400
BIC	409.6623	185.9500

Table 6 MANOVA results obtained on the real data partition and on that obtained by k-means and k-means-logistic regression

	Wilk's Lambda	Chi-Squared approximation	Degrees of freedom chi-squared	p-value	Partition
Const.	0.2052	267.6509	26	0.00E+00	Real
Group	0.7904	39.7581	12	7.89E−05	
Const.	0.2043	268.3793	26	0.00E+00	KM
Group	0.7609	44.3934	12	1.31E−05	
Const.	0.2303	248.1821	26	0.00E+00	KM-LR
Group	0.8063	36.3558	12	2.80E−06	

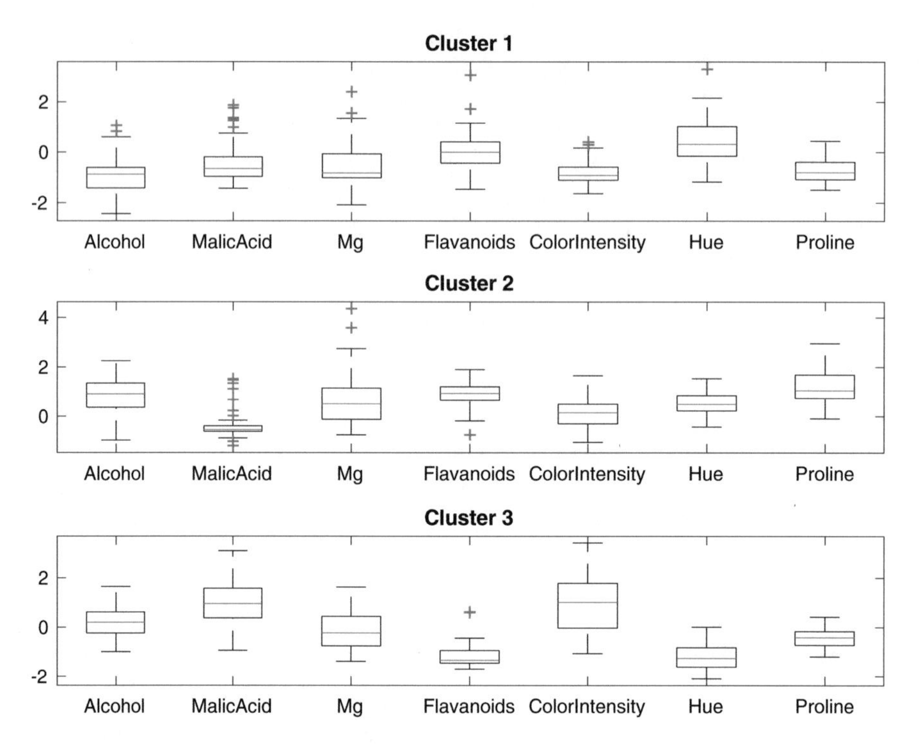

Fig. 2 Boxplots of the three KM-LR cluster distributions represented on the variables included in the model

4 Concluding Remarks

In the unsupervised classification approaches, the unknown number of clusters and the lack of assessment of the final partition are crucial issues that could negatively affect the reliability of the results. In this work we proposed an algorithm that combines KM and the LR modeling to evaluate the partition identified through KM, to assess the correct number of clusters, and to verify the selection of the most important variables. We did this by using well-known inferential tools that allowed us to statistically confirm the obtained results.

The application on real data shows that this methodology obtains better performance than the usual KM approach, reducing the effect of local minima. Moreover, KM-LR represents a useful tool to identify the variables that better contribute to defining the group structure in the data and removing the statistically non-significant variables from the model. In this way, we have a parsimonious set of variables that define the best partition of the data. Thus, the methodology seems promising. However, in a following work, we wish to better assess, using an extensive simulation study, the performance of the proposed methodology.

References

1. Agresti, A., Kateri, M.: Categorical data analysis. In: International Encyclopedia of Statistical Science, pp. 206–208 (2011)
2. Calinski, T., Harabasz, J.: A dendrite method for cluster analysis. Commun. Stat. Theory Methods **3**(1), 1–27 (1974)
3. Chaovalit, P., Zhou, L.: Movie review mining: a comparison between supervised and unsupervised classification approaches. In: Proceedings of the 38th Annual Hawaii International Conference on System Sciences, HICSS'05 (2005)
4. Dietterich, T.G.: Approximate statistical tests for comparing supervised classification learning algorithms. Neural Comput. **10**(7), 1895–1923 (1998)
5. Dua, D., Taniskidou, E.K.: UCI Machine Learning Repository. University of California, School of Information and Computer Science, Irvine (2017)
6. Everitt, B.S., Landau, S., Leese, M., Stahl, D.: Miscellaneous clustering methods. In: Cluster Analysis, 5th edn., pp. 215–255. Wiley, New York (2011)
7. Hastie, T., Tibshirani, R., Friedman, J.: Unsupervised learning. In: The Elements of Statistical Learning, pp. 485–585. Springer, Dordrecht (2009)
8. Hepner, G., Logan, T., Ritter, N., Bryant, N.: Artificial neural network classification using a minimal training set. Comparison to conventional supervised classification. Photogramm. Eng. Remote. Sens. **56**(4), 469–473 (1990)
9. Krzanowski, W.J., Lai, Y.T.: A criterion for determining the number of groups in a data set using sum of squares clustering. Biometrics **44**, 23–34 (1988)
10. MacQueen, J.: Some methods for classification and analysis of multivariate observations. In: Proceedings of the Fifth Berkeley Symposium on Mathematical Statistics and Probability **1**(14), 281–297 (1967)
11. Rand, W.M.: Objective criteria for the evaluation of clustering methods. J. Am. Stat. Assoc. **66**(336), 846–850 (1971)
12. Tibshirani, R., Walther, G., Hastie, T.: Estimating the number of clusters in a data set via the gap statistic. J. R. Stat. Soc. Ser. B Stat. Methodol. **63**(2), 411–423 (2001)

A Parametric Version of Probabilistic Distance Clustering

Christopher Rainey, Cristina Tortora, and Francesco Palumbo

Abstract Probabilistic distance (PD) clustering method grounds on the basic assumption that the product between the probability of the unit belonging to a cluster and the distance between the unit and the cluster center is constant, for each statistical unit. This constant is a measure of the classificability of the point, and the sum of the constant over units is referred to as the joint distance function (JDF). The parameters that minimize the JDF maximize the classificability of the units. The goal of this paper is to introduce a new distance measure based on a probability density function, specifically, we use the multivariate Gaussian and Student-t distributions. We show using two simulated data sets that the use of a distance based on these two density functions improves the performance of PD clustering.

Keywords PD clustering · Clustering algorithm · Gaussian distribution · Multivariate Student-t distribution

1 Introduction

Data clustering and classification are among the most investigated domains in statistics and machine learning. In general, clustering and classification methods can be divided into hierarchical and non-hierarchical methods. Non-hierarchical methods, the focus of this paper, produce a partition of the individuals into a specified number of groups by optimizing a numerical criterion [8]. Specifically, statistical non-hierarchical approaches are generally divided into two main categories: (1) heuristic

C. Rainey · C. Tortora
Department of Mathematics and Statistics, San José State University, San Jose, CA, USA
e-mail: cristina.tortora@sjsu.edu

F. Palumbo (✉)
Department of Political Sciences, University of Naples Federico II, Napoli, Italy
e-mail: fpalumbo@unina.it

F. Greselin et al. (eds.), *Statistical Learning of Complex Data*,
Studies in Classification, Data Analysis, and Knowledge Organization,
https://doi.org/10.1007/978-3-030-21140-0_4

(non-parametric) and (2) model-based approaches. The heuristic approach does not make any assumption about the class structure, and the criterion to optimize is generally based on a distance or dissimilarity measure. Class membership can be defined by a crisp function (clusters are mutually exclusive) or by a *fuzzy* function where a membership function is defined in [0, 1] for each unit and group, summing to 1 over groups, for each unit. Under this paradigm, the most used methods are k-means [12] and fuzzy c-means clustering [3]. The model-based approach postulates a formal statistical model for the classes—for example, that data were sampled from the Gaussian density—and it assumes that groups differ only by the density function parameter(s). Under this approach, the optimization problem consists in finding the parameters that maximize the likelihood function. Because the membership is unknown a common approach is to maximize the complete data likelihood, more details can be found in [13]. As the membership function is derived as a probability, it naturally varies in [0, 1] [6]. When the membership function is defined in [0, 1] the clustering approach is also defined as *probabilistic*.

This paper has a twofold aim: it builds a bridge between model-based and distance clustering by reformulating the probabilistic distance (PD) clustering algorithm, first introduced in 2008 by Ben-Israel and Iyigun in [2], using a density function; then, it proposes a probabilistic distance clustering algorithm using a Gaussian and a Student-t multivariate density distributions. Simulated data sets, using two different scenarios, have been used to show the algorithm performance.

The chapter is arranged into five sections including the present introduction. Section 2 briefly presents the PD clustering algorithm [2]; Sect. 3 reformulates the PD algorithm under a parametric paradigm; Sect. 4 presents an example on simulated data sets; and the last section provides some concluding remarks.

2 Probabilistic Distance Clustering

Probabilistic distance (PD) clustering was proposed by [2] in a distance based and distribution free context. It is a non-recursive, partitioning iterative algorithm, and it assumes that the number of clusters is known a priori. Given some random centers, the PD clustering algorithm assumes that the probability of a point belonging to a cluster is inversely proportional to the distance from the center of that cluster [10]. Suppose we have a data matrix $\mathbf{X}$ with n units and J variables, and consider K (non-empty) clusters. Let's denote by $\boldsymbol{x}_i$ a generic J-dimensional data vector and with $\mathbf{c}_k$ a general J-dimensional vector of centers, with $k = 1, \ldots, K, i = 1, \ldots, n$. PD clustering is based on two quantities: the distance of each $\boldsymbol{x}_i$ from each cluster center $\mathbf{c}_k$, denoted $d(\boldsymbol{x}_i, \mathbf{c}_k)$, and the probability of each point belonging to a cluster, i.e., $p(\boldsymbol{x}_i, \mathbf{c}_k)$, for $k = 1, \ldots, K$ and $i = 1, \ldots, n$. The relationship between them is the basic principle underlying the method.

For short, we define $p_{ik} := p(\boldsymbol{x}_i, \mathbf{c}_k)$ and $d_{ik} := d(\boldsymbol{x}_i, \mathbf{c}_k)$. PD clustering is based on the principle that the product of the distances and the probabilities is a constant depending only on $\boldsymbol{x}_i$ [2]. Denoting this constant by $F(\boldsymbol{x}_i)$, the following equality holds:

$$p_{ik}d_{ik} = F(\boldsymbol{x}_i), \tag{1}$$

where $F(\boldsymbol{x}_i)$ depends only on $\boldsymbol{x}_i$, i.e., $F(\boldsymbol{x}_i)$ does not depend on the cluster $k = 1, \ldots, K$. As the distance from the cluster center decreases, the probability of the point belonging to the cluster increases. Starting from (1), it is possible to compute p_{ik} as

$$p_{ik} = \frac{\prod_{m \neq k} d_{im}}{\sum_{m=1}^{K} \prod_{r \neq m} d_{ir}}, \qquad k = 1, \ldots K. \tag{2}$$

Then, from (1) and (2), it is possible to define the value of the constant $F(\boldsymbol{x}_i) = p_{ik}d_{ik}$ as

$$F(\boldsymbol{x}_i) = \frac{\prod_{m=1}^{K} d_{im}}{\sum_{m=1}^{K} \prod_{k \neq m} d_{ik}}, \qquad k = 1, \ldots K. \tag{3}$$

The quantity $F(\boldsymbol{x}_i)$ is a measure of the closeness of $\boldsymbol{x}_i$ to the cluster centers, and it determines the classificability of the point $\boldsymbol{x}_i$ with respect to the centers $\mathbf{c}_k$, for $k = 1, \ldots, K$. The smaller the $F(\boldsymbol{x}_i)$ value, the higher the probability of the point belonging to one cluster. If the distances between the point $\boldsymbol{x}_i$ and the centers of the clusters are all equal to d_i, then $F(\boldsymbol{x}_i) = d_i/K$ and all of the probabilities of belonging to each cluster are equal, i.e., $p_{ik} = 1/K$. The whole clustering problem consists of the identification of the centers that minimize the sum over i of $F(\boldsymbol{x}_i)$. In [2], the authors suggest using p^2 instead of p in (1) because it is a smoothed version of the problem, making the optimization function convex. The resulting quantity is called joint distance function (JDF):

$$\text{JDF} = \sum_{i=1}^{n} \sum_{k=1}^{K} d_{ik} p_{ik}^2. \tag{4}$$

Extensive details on PD clustering are given in [2]. In 2016, Tortora et al. proposed a factor version of the method to deal with high-dimensional data [18].

3 Methodology

A parametric version of the PD algorithm can be obtained using a dissimilarity measure based on a probability density function. Specifically, let's define with $M_k = \max(f(\boldsymbol{x}_i; \boldsymbol{\mu}_k, \boldsymbol{\theta}_k))$ the quantity $\log(M_k f(\boldsymbol{x}_i; \boldsymbol{\mu}_k, \boldsymbol{\theta}_k)^{-1})$ is a dissimilarity measure, where $f(\boldsymbol{x}_i; \boldsymbol{\mu}_k, \boldsymbol{\theta}_k)$ is a symmetric unimodal density function with location parameter $\boldsymbol{\mu}_k$ and parameter vector $\boldsymbol{\theta}_k$. Dealing with a generic density function $f(\boldsymbol{x}_i; \boldsymbol{\mu}_k, \boldsymbol{\theta}_k)$ one of the parameters is the mean, or alternatively it is the non-central parameter.

A general measure $d(x, y)$ is a dissimilarity measure if the following conditions are verified [16, p.404]:

1. $d(x, y) \geq 0$.
2. $d(x, y) = 0 \Leftrightarrow x = y$.
3. $d(x, y) = d(y, x)$.

Here we prove the following proposition.

Proposition *Let $f(\boldsymbol{x}_i; \boldsymbol{\mu}_k, \boldsymbol{\theta}_k)$ be the generic symmetric unimodal multivariate density function of the random variable X with parameter $\boldsymbol{\theta}_k$ and location parameter $\boldsymbol{\mu}_k$, then*

$$d(\boldsymbol{x}_i, \boldsymbol{\mu}_k) = \log\left(\frac{M_k}{f(\boldsymbol{x}_i; \boldsymbol{\mu}_k, \boldsymbol{\theta}_k)}\right) \tag{5}$$

satisfies all the three properties and it is a dissimilarity measure for $k = 1, \ldots, K$.

1. $d(\boldsymbol{x}_i, \boldsymbol{\mu}_k) > 0, \ \forall \boldsymbol{x}_i$.

Proof

$$0 < \frac{f(\boldsymbol{x}_i; \boldsymbol{\mu}_k, \boldsymbol{\theta}_k)}{M_k} \leq 1 \Rightarrow \frac{M_k}{f(\boldsymbol{x}_i; \boldsymbol{\mu}_k, \boldsymbol{\theta}_k)} \geq 1 \Rightarrow \log\left(\frac{M_k}{f(\boldsymbol{x}_i; \boldsymbol{\mu}_k, \boldsymbol{\theta}_k)}\right) \geq 0.$$

□

2. $d(\boldsymbol{x}_i, \boldsymbol{\mu}_k) = 0 \Leftrightarrow \boldsymbol{x}_i = \boldsymbol{\mu}_k$.
   2a. $\boldsymbol{x}_i = \boldsymbol{\mu}_k \Rightarrow d(\boldsymbol{x}_i, \boldsymbol{\mu}_k) = 0 \ \forall \boldsymbol{x}_i$.

Proof

$$\boldsymbol{x}_i = \boldsymbol{\mu}_k \Rightarrow f(\boldsymbol{x}_i; \boldsymbol{\mu}_k, \boldsymbol{\theta}_k) = f(\boldsymbol{\mu}_k; \boldsymbol{\mu}_k, \boldsymbol{\theta}_k) = M_k \Rightarrow \frac{M_k}{M_k} = 1 \Rightarrow \log(1) = 0.$$

□

2b. $d(\boldsymbol{x}_i, \boldsymbol{\mu}_k) = 0 \Rightarrow \boldsymbol{x}_i = \boldsymbol{\mu}_k, \ \forall \boldsymbol{x}_i$.

Proof

$$\log\left(\frac{M_k}{f(\boldsymbol{x}_i; \boldsymbol{\mu}_k, \boldsymbol{\theta}_k)}\right) = 0 \Rightarrow \frac{M_k}{f(\boldsymbol{x}_i; \boldsymbol{\mu}_k, \boldsymbol{\theta}_k)} = 1 \Rightarrow$$
$$\Rightarrow f(\boldsymbol{x}_i; \boldsymbol{\mu}_k, \boldsymbol{\theta}_k) = M_k = f(\boldsymbol{\mu}_k; \boldsymbol{\mu}_k, \boldsymbol{\theta}_k) \Rightarrow \boldsymbol{x}_i = \boldsymbol{\mu}_k.$$

□

3. $d(\boldsymbol{x}_i, \boldsymbol{\mu}_k) = d(\boldsymbol{\mu}_k, \boldsymbol{x}_i), \ \forall \boldsymbol{x}_i$.

Proof Given $\boldsymbol{\theta}_k$

$$f(\boldsymbol{x}_i; \boldsymbol{\mu}_k, \boldsymbol{\theta}_k) = f(\boldsymbol{\mu}_k; \boldsymbol{x}_i, \boldsymbol{\theta}_k), \Rightarrow \log\left(\frac{M_k}{f(\boldsymbol{x}_i;\boldsymbol{\mu}_k,\boldsymbol{\theta}_k)}\right) = \log\left(\frac{M_k}{f(\boldsymbol{\mu}_k;\boldsymbol{x}_i,\boldsymbol{\theta}_k)}\right).$$

□

Therefore, the quantity $\log\left(\frac{M_k}{f(\boldsymbol{x}_i; \boldsymbol{\mu}_k, \boldsymbol{\theta}_k)}\right)$ can be used in Eq. (4).

3.1 Gaussian PD Clustering

Gaussian PD clustering is obtained by putting

$$d_{ik} = \log\left(\frac{M_k}{\phi(\boldsymbol{x}_i; \boldsymbol{\mu}_k, \boldsymbol{\Sigma}_k)}\right)$$

in (4), where $\phi(\boldsymbol{x}_i; \boldsymbol{\mu}_k, \boldsymbol{\Sigma}_k)$ is the probability density function of a multivariate Gaussian distribution. In this case the loss function in Eq. (4) becomes

$$\begin{aligned} \text{JDF} = \textstyle\sum_{i=1}^{n}\sum_{k=1}^{K} p_{ik}^2 \log(M_k) + \sum_{i=1}^{n}\sum_{k=1}^{K} \frac{1}{2} p_{ik}^2 \log((2\pi)^J |\boldsymbol{\Sigma}_k|) \\ + \textstyle\sum_{i=1}^{n}\sum_{k=1}^{K} \frac{1}{2} p_{ik}^2 (\boldsymbol{x}_i - \boldsymbol{\mu}_k)' \boldsymbol{\Sigma}_k^{-1} (\boldsymbol{x}_i - \boldsymbol{\mu}_k). \end{aligned} \quad (6)$$

The parameters that minimize the objective function in (6) can be obtained by differentiating with respect to $\boldsymbol{\mu}_k$ and $\boldsymbol{\Sigma}_k$. Specifically, at a generic iteration *(t+1)*, the parameters that minimize (6) are

$$\boldsymbol{\mu}_k^{(t+1)} = \frac{\sum_{i=1}^{n} p_{ik}^2 \boldsymbol{x}_i}{\sum_{i=1}^{n} p_{ik}^2} \quad (7)$$

$$\boldsymbol{\Sigma}_k^{(t+1)} = \frac{\sum_{i=1}^{n} (\boldsymbol{x}_i - \boldsymbol{\mu}_k^{(t+1)})(\boldsymbol{x}_i - \boldsymbol{\mu}_k^{(t+1)})' p_{ik}^2}{\sum_{i=1}^{n} p_{ik}^2}. \quad (8)$$

Our iterative algorithm can be summarized as follows:

1. Random initialization of $\boldsymbol{\mu}_k$ and initialization of $\boldsymbol{\Sigma}_k$ as identity matrix, for $i = 1, \ldots, N$ and $k = 1, \ldots, K$;
2. update p_{ik} according to (2);
3. update $\boldsymbol{\mu}_k$ according to (7);
4. update $\boldsymbol{\Sigma}_k$ according to (8);
5. if $\boldsymbol{\mu}_k$ changed go to Step 2, otherwise stop.

In its parametric formalization for Gaussian (and Student-t) distributions, the JDF depends on both $\boldsymbol{\mu}_k$ and $\boldsymbol{\Sigma}_k$ (with $k = 1, \ldots, K$), then the quantity in (6) can be minimized for given values of $\boldsymbol{\Sigma}_k$; therefore, the convergence of the JDF to a minimum is not guaranteed. The same occurs for the PD clustering adjusted for cluster size algorithm [11], where the authors demonstrate that a convenient stopping rule is based on the stability of the solutions. In step 5, the algorithm stops when the difference in $\boldsymbol{\mu}_k$ from the previous step solution is negligible. It is worth noting that the quantity in (6) can be written as $\text{JDF} = \sum_{i=1}^{n}\sum_{k=1}^{K} p_{ik}^2(\log(M_k) - \log(\phi(\boldsymbol{x}_i; \boldsymbol{\mu}_k, \boldsymbol{\Sigma}_k)))$ with $M_k \geq \phi(\boldsymbol{x}_i; \boldsymbol{\mu}_k, \boldsymbol{\Sigma}_k)$; therefore, for every $k = 1, \ldots, K$ the function is upper-bounded for not degenerate density functions.

3.2 *Student-t PD Clustering*

Based on the same idea, we derived a Student-t PD clustering. In (6) the density function of a multivariate Gaussian distribution is replaced by the density function of a multivariate Student-t distribution,

$$f(\boldsymbol{x}, \boldsymbol{\mu}, \boldsymbol{\Sigma}, v) = \frac{\Gamma\left(\frac{v+J}{2}\right)|\boldsymbol{\Sigma}|^{-\frac{1}{2}}}{(\pi v)^{\frac{1}{2}J}\,\Gamma\left(\frac{v}{2}\right)\left\{1 + \frac{\delta(\boldsymbol{x},\boldsymbol{\mu},\boldsymbol{\Sigma})}{v}\right\}^{\frac{1}{2}(v+J)}}, \tag{9}$$

where $\delta(\boldsymbol{x}, \boldsymbol{\mu}, \boldsymbol{\Sigma}) = (\boldsymbol{x} - \boldsymbol{\mu})'\boldsymbol{\Sigma}^{-1}(\boldsymbol{x} - \boldsymbol{\mu})$. The JDF becomes

$$\text{JDF} = \sum_{i=1}^{n}\sum_{k=1}^{K} p_{ik}^2 \log(M_k) + \sum_{i=1}^{n}\sum_{k=1}^{K} p_{ik}^2\left[-\log\left(\Gamma\frac{v_k + J}{2}|\boldsymbol{\Sigma}_k|^{-\frac{1}{2}}\right)\right] \tag{10}$$

$$+\sum_{i=1}^{n}\sum_{k=1}^{K} p_{ik}^2\left[\log\left((\pi v_k)^{\frac{1}{2}J}\,\Gamma\left(\frac{v_k}{2}\right)\left(1 + \frac{\delta(\boldsymbol{x}_i, \boldsymbol{\mu}_k, \boldsymbol{\Sigma}_k)}{v_k}\right)^{\frac{1}{2}(v_k+J)}\right)\right].$$

The parameters that optimize Eq. (10) can be found by differentiating with respect to $\boldsymbol{\mu}_k$, $\boldsymbol{\Sigma}_k$, and v_k. Specifically, at a generic iteration *(t+1)*, the parameters that minimize the (10) are

$$\boldsymbol{\mu}_k^{(t+1)} = \frac{\sum_{i=1}^{n} w_{ik}\boldsymbol{x}_i}{\sum_{i=1}^{n} w_{ik}}, \tag{11}$$

$$w_{ik} = \frac{p_{ik}^2}{v_k^{(t)} + \delta\left(\boldsymbol{x}_i, \boldsymbol{\mu}_k^{(t)}, \boldsymbol{\Sigma}_k^{(t)}\right)}$$

$$\boldsymbol{\Sigma}_k^{(t+1)} = \frac{\sum_{i=1}^{n} p_{ik}^2(\boldsymbol{x}_i - \boldsymbol{\mu}_k^{(t+1)})(\boldsymbol{x}_i - \boldsymbol{\mu}_k^{(t+1)})'\frac{(v_k^{(t)}+J)}{v_k^{(t)}+\delta\left(\boldsymbol{x}_i,\boldsymbol{\mu}_k^{(t+1)},\boldsymbol{\Sigma}_k^{(t)}\right)}}{\sum_{i=1}^{n} p_{ik}^2}. \tag{12}$$

The degree of freedom $v_k^{(t+1)}$ is the solution to the following equation:

$$\sum_{i=1}^{n} p_{ik}^2 \left[\psi\left(\frac{v_k}{2}\right) - \psi\left(\frac{v_k + J}{2}\right) + \frac{J}{2v_k} + \frac{1}{2}\log\left(1 + \frac{\delta\left(\boldsymbol{x}_i, \boldsymbol{\mu}_k^{(t+1)}, \boldsymbol{\Sigma}_k^{(t+1)}\right)}{v_k^{(t)}}\right)\right]$$
$$-\frac{1}{2}\frac{v_k + J}{v_k} \sum_{i=1}^{n} p_{ik}^2 \frac{\delta\left(\boldsymbol{x}_i, \boldsymbol{\mu}_k^{(t+1)}, \boldsymbol{\Sigma}_k^{(t+1)}\right)}{v_k^{(t)} + \delta\left(\boldsymbol{x}_i, \boldsymbol{\mu}_k^{(t+1)}, \boldsymbol{\Sigma}_k^{(t+1)}\right)} = 0, \tag{13}$$

where $\psi(v) = \frac{\frac{\delta\Gamma(v)}{\delta v}}{\Gamma(v)}$.

Our iterative algorithm can be summarized as follows:

1. Random initialization of $\boldsymbol{\mu}_k$, initialization of $\boldsymbol{\Sigma}_k$ as identity matrix, and $v_k = 20$, for $i = 1, \ldots, N$ and $k = 1, \ldots, K$;
2. update p_{ik} according to (2);
3. update $\boldsymbol{\mu}_k$ according to (11);
4. update $\boldsymbol{\Sigma}_k$ according to (12);
5. update v_k solving (13);
6. if $\boldsymbol{\mu}_k$ changed, then go to Step 2, otherwise stop.

4 Application on Simulated Data Sets

Gaussian PD clustering (GPDC), Student-t PD clustering (TPDC), and standard PD clustering (PDC) algorithms have been compared on two simulated scenarios. For each scenario we generated 100 data sets, we set $k = 2$, $J = 2$, $n = 900$ ($n_1 = 400$, $n_2 = 500$), and we used the following parameters:

$$\boldsymbol{\mu}_1 = (0, 0)', \quad \boldsymbol{\mu}_2 = (2, 4)', \quad \boldsymbol{\Sigma}_1 = \begin{pmatrix} 1 & -0.5 \\ -0.5 & 1 \end{pmatrix}, \quad \text{and} \quad \boldsymbol{\Sigma}_2 = \begin{pmatrix} 1 & 0.5 \\ 0.5 & 1 \end{pmatrix}.$$

We used the software R [14], the standard PD clustering algorithm is fitted by the function `PDclust`, package `FPDclustering` [17]. For all the algorithms we used 5 random starts to find the starting points.

For scenario a, each cluster has been generated from a multivariate Gaussian distribution using the function `rmvnorm` from the R package `mvtnorm` [7]. For scenario b, each cluster has been generated from a multivariate Student-t distribution using the function `rmvt` from the same R package. Table 1 shows the true parameters compared with the average and the standard deviation of the estimated parameters. In Table 1, σ_{kij} refers to the elements of $\boldsymbol{\Sigma}_k$. In scenario a, all the methods give good estimates of the cluster means, and the parameters

Table 1 Mean and standard deviation (SD) of the parameters on 100 data sets

	Scenario a							Scenario b						
		GPDC		TPDC		PDC			GPDC		TPDC		PDC	
	True	Mean	SD	Mean	SD	Mean	SD	True	Mean	SD	Mean	SD	Mean	SD
μ_{11}	0.00	0.05	0.05	0.05	0.06	0.04	0.06	0.00	0.06	0.06	0.05	0.06	0.04	0.07
μ_{12}	0.00	−0.01	0.05	−0.01	0.05	0.11	0.06	0.00	0.02	0.07	0.01	0.06	0.13	0.07
μ_{21}	2.00	2.07	0.04	2.04	0.05	2.04	0.06	2.00	2.13	0.05	2.05	0.05	2.04	0.06
μ_{22}	4.00	4.07	0.04	4.04	0.05	4.03	0.07	4.00	4.13	0.05	4.04	0.05	4.03	0.06
σ_{111}	1.00	0.91	0.07	0.67	0.07			1.00	1.19	0.11	0.75	0.06		
σ_{112}	−0.50	−0.38	0.07	−0.26	0.05			−0.50	−0.49	0.11	−0.28	0.06		
σ_{122}	1.00	1.04	0.10	0.77	0.06			1.00	1.38	0.17	0.88	0.06		
σ_{211}	1.00	0.92	0.07	0.70	0.05			1.00	1.11	0.11	0.78	0.05		
σ_{212}	0.50	0.46	0.06	0.38	0.04			0.50	0.53	0.08	0.41	0.04		
σ_{222}	1.00	0.94	0.07	0.74	0.05			1.00	1.10	0.11	0.82	0.05		
v_1				5.97	0.02			8.00			6.05	1.05		
v_2				5.97	0.02			8.00			6.04	1.07		

obtained with TPDC are the closest to the true parameters. PDC doesn't estimate the covariance matrices of the clusters; both TPDC and GPDC give good estimates for the covariance matrices. It is worth noticing that TPDC estimates for σ are lower than the true parameters and lower than the estimates obtained with the GPDC. This can be explained by the estimates of the degrees of freedom that are smaller than the true values. In scenario b, all the methods give good estimates of the means; however, the estimates for the covariance matrices obtained using TPDC are closer to the true value when compared to GPDC. TPDC has an extra parameter compared to GPDC, the degrees of freedom, and this explains the higher variance for GPDC. The standard deviation (SD) of the mean estimates is similar for the three methods, the SD for the variance and covariance estimates for the GPDC is slightly higher, this difference decreases as the cluster dimensions increase.

To compare the clustering performance we use the adjusted Rand index (ARI) [9]. The ARI compares predicted with true classification, and corrects the Rand index [15] for chance; its expected value under random classification is 0, and it takes a value of 1 when there is perfect class agreement. In both scenarios the clusters overlap, see Fig. 1, despite that all the methods detect the clustering structure. Table 2 shows the average ARI values; the lowest ARI is 0.92 for scenario a and 0.86 for scenario b. TPDC outperforms the other methods in both scenarios with an ARI of 0.96 and 0.90, respectively. This is expected because TPDC is the most flexible method. The lower ARI of PD clustering is explained by its constant covariance structure. In Table 2 we also compare the proposed methods with k-means, Gaussian mixture models (GMM), and Student-t mixture models (TMM). For this comparison we used the R functions: `kmeans`, `gpcm` (option "VVV") package `mixture` [4], and `teigen` (option "UUUU") package `teigen` [1], respectively.

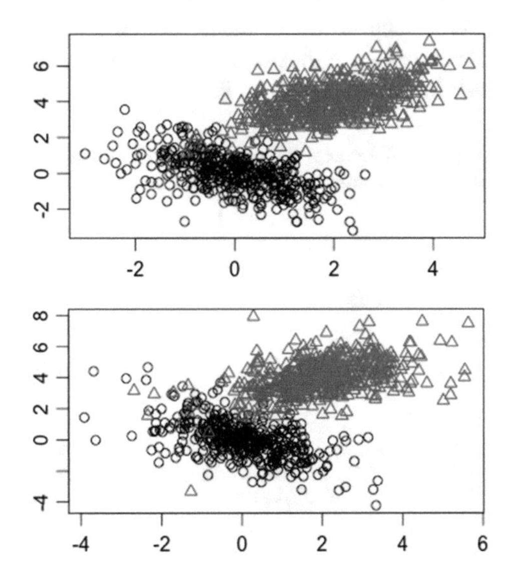

Fig. 1 Example of data sets generated using a mixture of multivariate Gaussian distributions (top) and a mixture of multivariate Student-t distributions with 8 degrees of freedom (bottom)

Table 2 Mean ARI and standard deviation (SD) on 100 data sets

	GPDC		TPDC		PDC		k-means		GMM		TMM	
	Mean	SD	Mean	SD	Mean	SD	Mean	SD	Mean	SD	Mean	SD
Scenario a	0.95	0.02	0.96	0.01	0.92	0.02	0.92	0.02	0.96	0.01	0.96	0.01
Scenario b	0.89	0.03	0.90	0.02	0.87	0.02	0.87	0.02	0.90	0.02	0.90	0.02

PD clustering and k-means give the same ARI in both scenarios. The data sets in scenario a and b have been generated as mixture of multivariate Gaussian and multivariate Student-t distributions, respectively; therefore, as expected, the two methods give the best performance, together with the proposed PD-t algorithm.

5 Conclusion and Future Work

In probabilistic distance (PD) clustering, given some random centers, the probability of a point belonging to a cluster is assumed to be inversely proportional to the distance from the center of that cluster. The algorithms perform very well; however, it shows limitations on clusters with difference variance or when variables are correlated. We proposed a parameterized version of the algorithm using a dissimilarity measure based on a probability density function. Specifically, we used the Gaussian and the Student-*t* distributions, this allows us to overcome the mentioned issues. Using simulated data sets, we showed that both algorithms can correctly estimate the parameters of the population and have great performance in terms of ARI. The same algorithm can be extended to other distributions; moreover, Iyigun and Ben-Israel [11] proposed the PD clustering algorithm adjusted for cluster size. A natural development of this work is to develop a parametric version of the PD clustering algorithm adjusted for cluster size and to compare it with the expectation–maximization algorithm [5].

Acknowledgements The authors are very grateful to the two anonymous referees for their detailed and helpful comments to finalize the manuscript.

References

1. Andrews, J.L., Wickins, J.R., Boers, N.M., McNicholas, P.D.: teigen: an R package for model-based clustering and classification via the multivariate *t* distribution. J. Stat. Softw. **83**, 1–32 (2017)
2. Ben-Israel, A., Iyigun, C.: Probabilistic d-clustering. J. Classif. **25**, 5–26 (2008)
3. Bezdek, J.C., Ehrlich, R., Full, W.: FCM: the fuzzy c-means clustering algorithm. Comput. Geosci. **10**, 191–203 (1984)

4. Browne, R.P., ElSherbiny, A., McNicholas, P.D.: FCM: mixture: Mixture Models for Clustering and Classification. R package version 1.4 (2015). https://cran.r-project.org/web/packages/mixture/index.html
5. Dempster, A.P., Laird, N.M., Rubin, D.B.: Maximum likelihood from incomplete data via the EM algorithm. J. R. Stat. Soc. B-met Ser. B **39**, 1–38 (1977)
6. Everitt, B.S., Landau, S., Leese, M., Stahl, D.: Cluster Analysis. Wiley Series in Probability and Statistics. Wiley, New York (2011)
7. Genz, A., Bretz, F., Miwa, T., Mi, X., Leisch, F., Scheipl, F., Hothorn, T.: mvtnorm: multivariate normal and *t* distributions. R package version 1.0-7 (2009). https://cran.r-project.org/web/packages/mvtnorm/index.html
8. Gordon, A.D.: Classification, 2nd edn. Chapman and Hall/CRC, Boca Raton (1999)
9. Hubert, L., Arabie, P.: Comparing partitions. J. Classif. **2**, 193–218 (1985)
10. Iyigun, C.: Probabilistic distance clustering. Ph.D. thesis, State University of New Jersey (2007)
11. Iyigun, C., Ben-Israel, A.: Probabilistic distance clustering adjusted for cluster size. Probab. Eng. Inform. Sci. **22**, 68–125 (2008)
12. MacQueen, J.: Some methods for classification and analysis of multivariate observations. In: Proceedings of the Fifth Berkeley Symposium, vol. 1, pp. 281–297 (1967)
13. McLachlan, G.J., Peel, D.: Finite Mixture Models. Wiley Interscience, New York (2000)
14. R Core Team: R: a language and environment for statistical computing. R Foundation for Statistical Computing, Vienna (2016)
15. Rand, W.M.: Objective criteria for the evaluation of clustering methods. J. Am. Stat. Assoc. **66**, 846–850 (1971)
16. Theodoridis, S., Koutroumbas, K.: Pattern Recognition, 2nd edn. Academic Press, New York (2003)
17. Tortora, C., McNicholas, P.D.: FPDclustering: PD-clustering and factor PD-clustering. R package version 1.1 (2016). https://cran.r-project.org/web/packages/FPDclustering/index.html
18. Tortora, C., Gettler-Summa, M., Marino, M., Palumbo, F.: Factor probabilistic distance clustering (FPDC): a new clustering method. Adv. Data Anal. Classif. **10**, 441–464 (2016)

An Overview on the URV Model-Based Approach to Cluster Mixed-Type Data

Monia Ranalli and Roberto Rocci

Abstract In this paper, we provide an overview on the underlying response variable (URV) model-based approach to cluster and, optionally, simultaneously reduce ordinal and, optionally, continuous variables. We summarize and compare its main features discussing some key issues. An example of application to real data is illustrated comparing and discussing clustering performances.

Keywords URV · Finite mixture models · Ordinal data · Composite likelihood

1 Introduction

A frequently used clustering model is the finite mixture of Gaussians (FMG) [15],

$$f(\mathbf{y};\boldsymbol{\theta}) = \sum_{g=1}^{G} p_g \phi_P\left(\mathbf{y};\boldsymbol{\mu}_g,\boldsymbol{\Sigma}_g\right), \tag{1}$$

where $\phi_P\left(\mathbf{y};\boldsymbol{\mu}_g,\boldsymbol{\Sigma}_g\right)$ is the P-variate Gaussian density with mean $\boldsymbol{\mu}_g$ and covariance matrix $\boldsymbol{\Sigma}_g$ and $p_1, p_2, \ldots, p_G$ is the set of positive weights that sum to 1. Usually each Gaussian density (component) is interpreted as a cluster (sub-population) and the corresponding weight as the probability that an observation comes from it. FMG works on continuous variables, but some issues arise on ranked data due to the lack of metric properties: the category scores are arbitrary and the assumption of normality is not true anymore. To analyse ordinal data two main approaches exist: item response theory (IRT, [1]) and underlying response variable (URV, [19]). In the first one, the ordinal variables are assumed to be independent given a set of latent continuous variables that have a clustering structure (for

M. Ranalli (✉) · R. Rocci
Department of Economics and Finance, University of Tor Vergata, Rome, Italy
e-mail: monia.ranalli@uniroma2.it; roberto.rocci@uniroma2.it

F. Greselin et al. (eds.), *Statistical Learning of Complex Data*,
Studies in Classification, Data Analysis, and Knowledge Organization,
https://doi.org/10.1007/978-3-030-21140-0_5

example, they can be distributed as a FMG [4]). On the other hand, URV is a way to overcome the within-independence limitation: the observed variables are a categorization of underlying non-observable continuous variables distributed as a FMG (see, for example, [8, 12, 20, 22, 23]). This has been extended in several ways. Everitt [8] introduces a mixture model for mixed data. The joint distribution of the variables is a homoscedastic FMG where some variables are observed as ordinal. In particular, the ordinal variables are seen as generated by thresholding some marginals of the joint FMG with different thresholds in each component. The model proposed by Lubke and Neale [12] is specified for ordinal variables that are generated by thresholding a heteroscedastic mixture of Gaussians, whose covariance matrices are reparametrized as a factor analysis model. Nevertheless, in both cases the estimation of the model by maximum likelihood requires the numerical computation of multidimensional integrals that is time consuming. Due to computational reasons, they can include only few ordinal variables. In the sequel we summarize the main results obtained by adopting a composite likelihood approach. We first present the model with only ordinal variables [20], then we extend it to the case where noise dimensions or variables are present [23] and finally we generalize the proposal to the mixed-type data [22].

2 Clustering Ordinal Data

We start by describing the key figures for the proposal of [20]. This aims at capturing the cluster structure underlying the data without requiring the local independence assumption that may result to be too restrictive in practice. Let $x_1, \ldots, x_P$ be ordinal variables and $c_i = 1, \ldots, C_i$ the associated categories for $i = 1, \ldots, P$. There are $R = \prod_{i=1}^{P} C_i$ possible response patterns $\mathbf{x}_r = (x_1 = c_1, \ldots, x_P = c_P)$, with $r = 1, \ldots, R$. The ordinal variables are generated by thresholding $\mathbf{y}$ that is a multivariate continuous random variable distributed as a FMG (1). The link between $\mathbf{x}$ and $\mathbf{y}$ is expressed by a threshold model defined as $x_i = c_i \Leftrightarrow \gamma^{(i)}_{c_i-1} \leq y_i < \gamma^{(i)}_{c_i}$. Let $\boldsymbol{\psi} = \left\{p_g, \ldots, p_{G-1}, \boldsymbol{\mu}_1, \ldots, \boldsymbol{\mu}_G, \boldsymbol{\Sigma}_1, \ldots, \boldsymbol{\Sigma}_G, \boldsymbol{\Gamma}\right\}$ be the set of model parameters, where Γ is the set of vectors $\boldsymbol{\gamma}^{(i)}$. The probability of response pattern $\mathbf{x}_r$ is

$$\Pr(\mathbf{x}_r; \boldsymbol{\psi}) = \sum_{g=1}^{G} p_g \int_{\gamma^{(1)}_{c_1-1}}^{\gamma^{(1)}_{c_1}} \cdots \int_{\gamma^{(P)}_{c_P-1}}^{\gamma^{(P)}_{c_P}} \phi(\mathbf{y}; \boldsymbol{\mu}_g, \boldsymbol{\Sigma}_g) d\mathbf{y} = \sum_{g=1}^{G} p_g \pi_r(\boldsymbol{\mu}_g, \boldsymbol{\Sigma}_g, \boldsymbol{\Gamma}), \tag{2}$$

where $\pi_r(\boldsymbol{\mu}_g, \boldsymbol{\Sigma}_g, \boldsymbol{\Gamma})$ is the probability of response pattern $\mathbf{x}_r$ in cluster g. Thus, for a random i.i.d. sample of size N the log-likelihood is

$$\ell(\boldsymbol{\psi}; \mathbf{X}) = \sum_{r=1}^{R} n_r \log \left[\sum_{g=1}^{G} p_g \pi_r \left(\boldsymbol{\mu}_g, \boldsymbol{\Sigma}_g, \boldsymbol{\Gamma} \right) \right], \tag{3}$$

where n_r is the observed sample frequency of response pattern $\mathbf{x}_r$ and $\sum_{r=1}^{R} n_r = N$. The maximization of (3) is quite time consuming and becomes infeasible when P increases due to the presence of multidimensional integrals. For this reason, model parameters are estimated through an EM framework maximizing the pairwise log-likelihood, i.e. the sum of all possible log-likelihoods based on the bivariate marginals. The estimators obtained have been proven to be consistent, asymptotically unbiased and normally distributed. In general they are less efficient than the full maximum likelihood estimators, but in many cases the loss in efficiency is very small or almost null [25].

3 Simultaneous Clustering and Reduction

In order to identify the discriminative dimensions, in the previously described model, Ranalli and Rocci [23] assumed that there is a second order set of P latent variables, say factors, $\tilde{\mathbf{y}}$, formed of two independent subsets. In the first one, there are Q (with $Q < P$) factors that have some clustering information, defining the so-called discriminative dimensions. In the second set there are $\bar{Q} = P - Q$ noise factors, i.e. noise dimensions. Technically, the Q informative elements of $\tilde{\mathbf{y}}$ are assumed to be distributed as a mixture of Gaussians with class conditional means and variances equal to $E(\tilde{\mathbf{y}}^Q \mid g) = \boldsymbol{\eta}_g$ and $\mathrm{Cov}(\tilde{\mathbf{y}}^Q \mid g) = \boldsymbol{\Omega}_g$, respectively. The $\bar{Q}$ noisy elements do not contain information about the cluster structure, it follows that they are independent of $\tilde{\mathbf{y}}^Q$ and their distribution does not vary from one class to another: $E(\tilde{\mathbf{y}}^{\bar{Q}} \mid g) = \boldsymbol{\eta}_0$, $\mathrm{Cov}(\tilde{\mathbf{y}}^{\bar{Q}} \mid g) = \boldsymbol{\Omega}_0$. The link between the latent variables and the factors is given by $\mathbf{y} = \mathbf{A}\tilde{\mathbf{y}}$, where $\mathbf{A}$ is non-singular. The variables $\mathbf{y}$ that are most correlated with the factors $\tilde{\mathbf{y}}^{\bar{Q}}$ are identified as noise. It is also worth noticing that, exploiting the independence between $\tilde{\mathbf{y}}^Q$ and $\tilde{\mathbf{y}}^{\bar{Q}}$, it is possible to compute proportions of each latent variable's variance that can be explained by the noise factors and, by one's complement, by the discriminative factors. They are very helpful in identifying the noise variables. For more details, see [23].

4 Clustering Mixed-Type Data

Finally, we summarize the extension of [20] to the mixed-type data case [22] (called as "HetMixtureMixed"). Let $\mathbf{x} = [x_1, \ldots, x_Q]'$ and $\mathbf{y}^{\bar{Q}} = [y_{Q+1}, \ldots, y_P]'$ be Q ordinal and $\bar{Q} = P - Q$ continuous variables, respectively. Under the URV, the ordinal variables $\mathbf{x}$ are considered as a discretization of a continuous multivariate latent variable $\mathbf{y}^Q = [y_1, \ldots, y_Q]'$. To accommodate both cluster structure and dependence within the groups, we assume that $\mathbf{y} = [\mathbf{y}^{Q'}, \mathbf{y}^{\bar{Q}'}]'$ follows the heteroscedastic Gaussian mixture (1). For a random i.i.d. sample of size N, $(\mathbf{x}_1, \mathbf{y}_1^{\bar{Q}}), \ldots, (\mathbf{x}_N, \mathbf{y}_N^{\bar{Q}})$, the log-likelihood is

$$\ell(\boldsymbol{\psi}) = \sum_{n=1}^{N} \log \left[\sum_{g=1}^{G} p_g \phi_{\bar{Q}}(\mathbf{y}_n^{\bar{Q}}; \boldsymbol{\mu}_g^{\bar{Q}}, \boldsymbol{\Sigma}_g^{\bar{Q}}) \pi_n \left(\boldsymbol{\mu}_{n;g}^{Q|\bar{Q}}, \boldsymbol{\Sigma}_g^{Q|\bar{Q}}, \boldsymbol{\Gamma} \right) \right],$$

where $\pi_n \left(\boldsymbol{\mu}_{n;g}^{Q|\bar{Q}}, \boldsymbol{\Sigma}_g^{Q|\bar{Q}}, \boldsymbol{\Gamma} \right) = \int_{\gamma_{c_1-1}^{(1)}}^{\gamma_{c_1}^{(1)}} \cdots \int_{\gamma_{c_Q-1}^{(Q)}}^{\gamma_{c_Q}^{(Q)}} \phi_Q(\mathbf{u}; \boldsymbol{\mu}_{n;g}^{Q|\bar{Q}}, \boldsymbol{\Sigma}_g^{Q|\bar{Q}}) d\mathbf{u}$ that is the conditional joint probability of response pattern $\mathbf{x}_n = (c_1^{(1)}, \ldots, c_Q^{(Q)})$ given the cluster g and the continuous variables $\mathbf{y}_n^{\bar{Q}}$. To overcome the computational issues caused by the presence of multidimensional integrals in the likelihood, a composite likelihood is used, composed of three block-estimating functions: the full likelihood of a FMG for the continuous variables, the pairwise likelihood of a latent mixture of Gaussians for the ordinal variables ($Q(Q-1)/2$ sub-likelihoods) and the Q likelihoods of one ordinal variable and all continuous variables. For more details, see [22].

5 Model Identifiability

The combination of ordinal data and composite likelihood requires specific attention to identifiability issues. Composite likelihood estimation methods provide good estimators as long as the model is identified, i.e. if the composite likelihood is rich enough to include all the information about the parameters [14]. In other words, the marginals involved in the composite likelihood should be able to capture and identify the true cluster structure underlying the data. As an example, an identified model could be not identified looking only at all the bivariate marginals. We illustrate this aspect through an example for continuous data. The first row of Fig. 1 displays two different cluster structures, both generated from a tri-variate homoscedastic FMG with four components equally weighted. The only difference is given by the centroids of the clusters. The remaining rows show that the same bivariate marginals correspond to two different configurations of four clusters. It follows that, in some cases, it is not possible to identify the true cluster structure

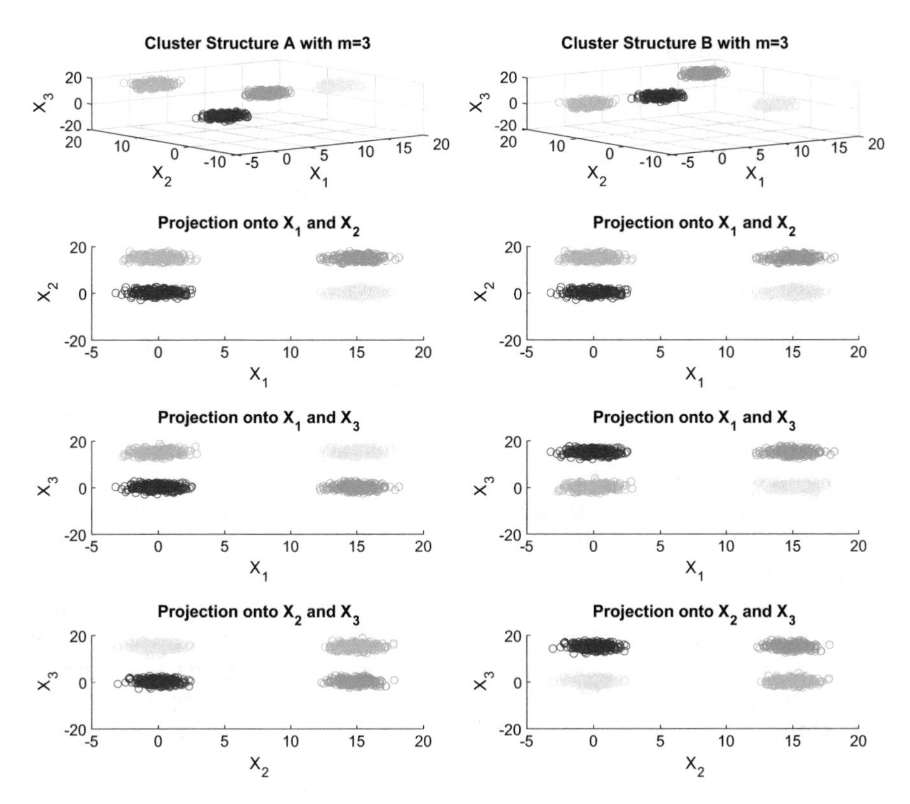

Fig. 1 Simulated example where true cluster structure can be captured only with $m = 3$. Two different cluster structures (first row) lead to the same bivariate marginals (last three rows)

by looking at only the bivariate marginals. However, we note that in the previous example the non-identifiability is due to the perfect overlapping of the centroids of the clusters on the marginals, in addition to the same covariance matrix and mixture weights. In practice, these conditions are strict and very unlikely because they are based on several equalities of parameters. Summing up, we believe that situations of non-identifiability are rare in practice, especially with a large number of variables and the real problem is the case where the model is weakly identifiable (see [23] for further details). In such cases, it is recommended to use higher marginal orders. This leads to increase the efficiency of the composite estimators and to improve the model identifiability. Specific details and some necessary/sufficient conditions can be found in [20, 22, 23].

6 Computation, Classification and Model Selection

All the models summarized above are estimated within the expectation–maximization (EM) framework maximizing a composite log-likelihood. As regards the classification, in the context of finite mixture models estimated through a full likelihood, an observation is assigned to the component with the maximum a posteriori probability (MAP criterion). However, when we adopt a composite likelihood approach, this is not possible anymore, since we do not compute the joint density for each observation. To solve the problem there are at least two different solutions [22, 23]. In the first, the MAP criterion is used where the joint probabilities are estimated by evaluating the multidimensional integrals on the composite-estimates. In the second, we note that the MAP criterion assigns an observation to the component with the maximum scaled fit (scaled by the corresponding mixing weight). Similarly, in the composite likelihood framework: for one observation it is evaluated its (scaled) composite fit on each component and it is assigned to the component corresponding to the maximum (scaled) fit (CMAP criterion). In the first case, it is true that there are still multidimensional integrals, but they have not to be evaluated many times (as it is needed in the estimation), but only once. However, CMAP is more efficient computationally, with competitive performance, as shown in [22]. Finally, the best model is chosen by minimizing the composite version of penalized likelihood selection criteria like BIC or CLC (see [21] and the references therein).

7 Some Related Models

The aforementioned models could be seen as an extension, with some modifications, of Everitt's proposal [8] into the composite estimation framework. It would be also interesting to make such extension for the proposal of [12]. On the other hand, the proposal of [23] can be compared to variable selection and parsimonious modelling. Variable selection (see, e.g. [7, 27] for categorical data) is commonly based on heuristic methods that are computationally demanding. It assumes that only noise variables may exist—it is not assumed the existence of noisy dimensions. The proposal of [23] can be used to understand how much a variable is informative or not for the classification. Within the second purpose, examples of parsimonious modelling in the context of continuous data are [6], mixtures of factor analysers (see, e.g. [16, 17]), mixtures of principal component analysers (see, e.g. [24]). See [2] for a recent review on model-based clustering of high-dimensional data. As regards categorical data, we find few analogous proposals (see, e.g. [9, 12, 13]). All the aforementioned proposals use a variable reduction model, like factor analysis or principal component analysis, to reparametrize the component covariance matrices. They do not aim at identifying the informative dimensions. Their dimensionality reduction is only local and within the components.

Differently, [23] can be used to cluster observations by taking into account the presence of global informative/noise dimensions. Finally [22] can be compared with some existing proposals on clustering mixed-type data. Lawrence and Krzanowski [11] introduced a location mixture model according to which for the continuous variables, a Gaussian mixture exists, whose component mean vectors depend on the specific combination of categories (i.e. response pattern) assumed by the categorical ones. However, it is not identifiable without imposing some constraints on the mean parameters of the Gaussian distributions [28]. Furthermore each combination of categories identifies a set of clusters: it follows that the total number of clusters can be unnecessarily large. A more parsimonious, but less realistic, model is given by Hunt and Jorgensen [10], according to which the variables are decomposed into conditionally independent blocks containing a set of continuous variables or one categorical variable. A more general model can be obtained by exploiting the IRT approach. The observed variables, continuous or ordinal, are assumed to be independent given a set of continuous latent variables that have a clustering structure described by a FMG [3, 5]. Finally, by relaxing the local independence assumption, Morlini [18] proposes a model-based clustering for mixed binary and continuous variables. The estimation is carried out in two steps using the software LATENT GOLD [26]. Differently from the location mixture model proposed by Lawrence and Krzanowski [11] and the model proposed by Hunt and Jorgensen [10], in [22] there is no local independence or conditionally independent blocks assumption. Differently from [3], in [22] the dependencies between variables, both within and between groups, can be easily measured. Differently from [18], in [22] the parameter estimates are carried out simultaneously.

8 Real Data Application

Data is composed of 1599 Portuguese "Vinho Verde" wine (red wine) described by eleven physicochemical continuous variables (fixed acidity, volatile acidity, citric acidity, residual sugar, chlorides, free sulphur dioxide, total density dioxide, density, pH, sulphates, alcohol) and one ordinal variable (note of quality). Different models have been compared: the FMG for all data (naive approach—treating the ordinal variable as continuous), the latent FMG only for the ordinal variable and the model illustrated in Sect. 4 (HetMixtureMixed). All models have been fitted with different number of groups, $G = 2, 3, 4$. The cCLC (144,990) selects $G = 3$ as the best solution for the HetMixtureMixed model, compared to 181,800 and 154,660 for $G = 2$ and $G = 4$, respectively. The BIC selects $G = 2$ as the best solution for FMG on all data (naive approach) and the latent FMG on the only ordinal variable. For the naive approach the BIC is 16,547, compared to 19,185 and 19,645 for $G = 3$ and $G = 4$, respectively. For the latter the BIC is 16,547 compared to 19,185 and 19,645 for $G = 3$ and $G = 4$, respectively. The difference in G across the three models can be justified as follows: the presence of the ordinal variable tends to guide the choice of cluster number. Indeed, looking at the relative frequency distribution of the ordinal variable, the number of groups tends to coincide with the number of categories most frequent (quality variable assumes value $k = 1, \ldots, 6$ with the following frequencies 0.01, 0.03, 0.43, 0.40, 0.12, 0.01, respectively). Furthermore, the adjusted Rand index obtained comparing the fitted partition provided by the latent FMG only for the ordinal variable with the partition obtained by assigning to the wines label 1 if $x_k = 1, 2, 3$ and 2 otherwise is equal to 0.8485. It follows that by fitting the latent FMG for mixed-type data (and thus by taking into account the nature of ordinal variables properly), it mitigates the effect of the ranks on the clusters. The three groups fitted by the latent FMG for mixed-type data represent different values of wine quality: high quality ($p_1 = 0.61$), medium quality ($p_2 = 0.22$) and low quality ($p_3 = 0.17$). The high quality wine group is characterized mainly by lower levels of acidity, pH, chlorides and sulphites (these levels will increase as the wine quality decreases). It presents high correlation between both sulphur measures opposite to a high correlation between the density and acidity measures. The low quality wine group takes larger values for both sulphur dioxide measures and the alcoholic rate. In this class, the wine quality is correlated with a large alcoholic measure and small values for the chlorides and acidity measures. The second and the third group present similar features; the main difference is given by the total sulphur dioxide that is twice in group 2 than in group 3.

References

1. Bock, D., Moustaki, I.: Item response theory in a general framework. In: Handbook of Statistics on Psychometrics. Elsevier, Amsterdam (2007)
2. Bouveyron, C., Brunet, C.: Model-based clustering of high-dimensional data: a review. Comput. Stat. Data Anal. **71**, 52–78 (2012)
3. Browne, R.P., McNicholas, P.D.: Model-based clustering, classification, and discriminant analysis of data with mixed type. J. Stat. Plan. Inference **142**(11), 2976–2984 (2012)
4. Cagnone, S., Viroli, C.: A factor mixture analysis model for multivariate binary data. Stat. Model. **12**, 257–277 (2012)
5. Cai, J.H., Song, X.Y., Lam, K.H., Ip, E.H.S.: A mixture of generalized latent variable models for mixed mode and heterogeneous data. Comput. Stat. Data Anal. **55**(11), 2889–2907 (2011)
6. Celeux, G., Govaert, G.: Gaussian parsimonious clustering models. Pattern Recognit. **28**(5), 781–793 (1995)
7. Dean, N., Raftery, A.E.: Latent class analysis variable selection. Ann. Inst. Stat. Math. **62**(1), 11–35 (2010)
8. Everitt, B.: A finite mixture model for the clustering of mixed-mode data. Stat. Probab. Lett. **6**(5), 305–309 (1988)
9. Gollini, I., Murphy, T.: Mixture of latent trait analyzers for model-based clustering of categorical data. Stat. Comput. **24**(4), 569–588 (2014)
10. Hunt, L., Jorgensen, M.: Clustering mixed data. Wiley Interdiscip. Rev. Data Min. Knowl. Discov. **1**(4), 352–361 (2011)
11. Lawrence, C., Krzanowski, W.: Mixture separation for mixed-mode data. Stat. Comput. **6**(1), 85–92 (1996)
12. Lubke, G., Neale, M.: Distinguishing between latent classes and continuous factors with categorical outcomes: class invariance of parameters of factor mixture models. Multivar. Behav. Res. **43**(4), 592–620 (2008)

13. Marbac, M., Biernacki, C., Vandewalle, V.: Finite mixture model of conditional dependencies modes to cluster categorical data (2014, preprint). arXiv:1402.5103
14. Mardia, K.V., Kent, J.T., Hughes, G., Taylor, C.C.: Maximum likelihood estimation using composite likelihoods for closed exponential families. Biometrika **96**(4), 975–982 (2009)
15. McLachlan, G.J., Rathnayake, S.I.: Mixture models for standard p-dimensional Euclidean data. In: Hennig, C., Meila, M., Murtagh, F., Rocci, R. (eds.) Handbook of Cluster Analysis, pp. 145–172. CRC Press, Boca Raton (2016)
16. McLachlan, G.J., Bean, R.W., Ben-Tovim Jones, L.: Extension of the mixture of factor analyzers model to incorporate the multivariate t-distribution. Comput. Stat. Data Anal. **51**, 5327–5338 (2007)
17. McNicholas, P., Murphy, T.: Parsimonious Gaussian mixture models. Stat. Comput. **18**(3), 285–296 (2008)
18. Morlini, I.: A latent variables approach for clustering mixed binary and continuous variables within a Gaussian mixture model. Adv. Data Anal. Classif. **6**(1), 5–28 (2012)
19. Muthén, B.: A general structural equation model with dichotomous, ordered categorical, and continuous latent variable indicators. Psychometrika **49**(1), 115–132 (1984)
20. Ranalli, M., Rocci, R.: Mixture models for ordinal data: a pairwise likelihood approach. Stat. Comput. **26**(1), 529–547 (2016)
21. Ranalli, M., Rocci, R.: Standard and novel model selection criteria in the pairwise likelihood estimation of a mixture model for ordinal data. In: Wilhelm, A.F.X., Kestler, H.A. (eds.) Analysis of Large and Complex Data. Studies in Classification, Data Analysis and Knowledge Organization, pp. 53–68. Springer, Cham (2016)
22. Ranalli, M., Rocci, R.: Mixture models for mixed-type data through a composite likelihood approach. Comput. Stat. Data Anal. **110**(C), 87–102 (2017). https://doi.org/10.1016/j.csda.2016.12.01
23. Ranalli, M., Rocci, R.: A model-based approach to simultaneous clustering and dimensional reduction of ordinal data. Psychometrika (2017). https://doi.org/10.1007/s11336-017-9578-5
24. Tipping, M.E.: Probabilistic visualisation of high-dimensional binary data. In: Proceedings of the 1998 Conference on Advances in Neural Information Processing Systems II, pp. 592–598. MIT Press (1999)
25. Varin, C., Reid, N., Firth, D.: An overview of composite likelihood methods. Stat. Sin. **21**(1), 1–41 (2011)
26. Vermunt, J.K., Magidson, J.: Latent GOLD 4.0 User's Guide. Statistical Innovations Inc., Belmont (2005)
27. White, A., Wyse, J., Murphy, T.B.: Bayesian variable selection for latent class analysis using a collapsed Gibbs sampler (2014, preprint). arXiv:1402.6928
28. Willse, A., Boik, R.: Identifiable finite mixtures of location models for clustering mixed-mode data. Stat. Comput. **9**(2), 111–121 (1999)

Part II
Exploratory Data Analysis

Preference Analysis of Architectural Façades by Multidimensional Scaling and Unfolding

Giuseppe Bove, Nicole Ruta, and Stefano Mastandrea

Abstract The methods of paired comparison and ranking play an important role in the analysis of preference data. In this study, first we show how asymmetric multidimensional scaling allows to represent in a diagram the preference order that comes out in a paired-comparison task concerning architectural façades. A ranking task involving the same stimuli and the same subject sample further enriched the preference analysis, because multidimensional unfolding applied to the ranking data matrix allows to detect the relationships between subjects and architectural façades. The results show that high curved façade is the most preferred, followed by the medium curved, angular and rectilinear ones. Rectilinear stimuli were always the least preferred and not angularity as expected.

Keywords Preference data · Asymmetric multidimensional scaling · Multidimensional unfolding

1 Introduction

Several studies showed that people prefer curved objects compared to angular ones and that curved polygons are more easily associated with safe and positive concepts and with female names compared to their angular counterpart, but the elements that drive this preference are still unclear [7]. In this study, the role of curvature in driving preferences is generalized to the architecture domain, by

G. Bove (✉) · S. Mastandrea
Department of Education, Roma Tre University, Rome, Italy
e-mail: giuseppe.bove@uniroma3.it; stefano.mastandrea@uniroma3.it

N. Ruta
Cardiff Metropolitan University, School of Art and Design, Cardiff, UK
e-mail: niruta@cardiffmet.ac.uk

F. Greselin et al. (eds.), *Statistical Learning of Complex Data*,
Studies in Classification, Data Analysis, and Knowledge Organization,
https://doi.org/10.1007/978-3-030-21140-0_6

focusing on classical architectural façades [4]. The classical *Oratorio dei Filippini* architecture by Francesco Borromini (Rome, 1637–1650) was chosen as reference building, a typical Baroque building due to its characteristic curved lines. We asked a professional architect to render a simplified 2D model of the selected architectural façade in order to make our stimuli more realistic. The architectural façade was modified controlling for the global and local amount of curvature to introduce. Four versions of the same building were produced and used in the experiment (see Fig. 1): a. high curvature: global and local curvature; b. medium curvature: global curvature and local straight; c. rectilinear: global and local straight; d. angular: global and local angular.

Preferences were collected from twenty-four volunteers recruited from the student population of Roma Tre University, with two different methods: a paired-comparison task and a ranking task. To further investigate the role of expertise, at the end of the tasks participants were asked to self-report on a five-point Likert scale their artistic education level and their art interest.

In the following sections we will report separately the results of the paired-comparison and ranking tasks. Conclusions are discussed in the final section.

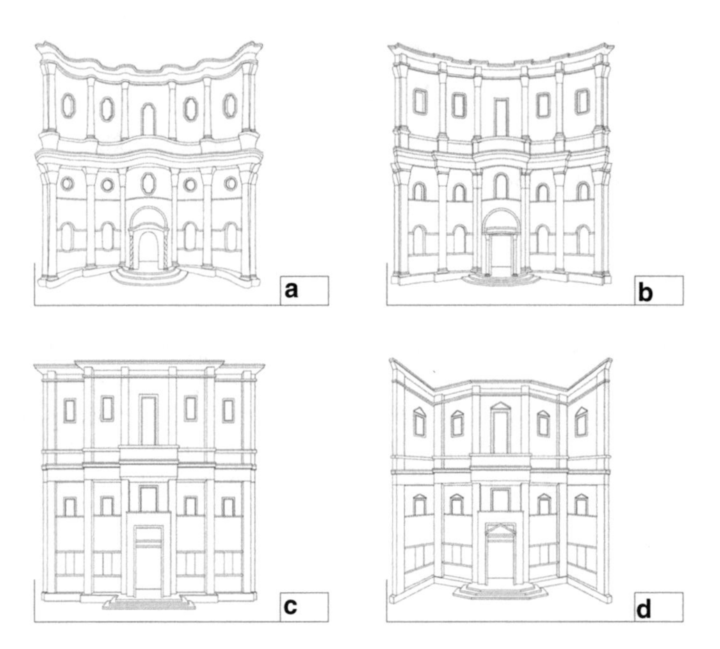

Fig. 1 The four architectural façades used in the study. From (**a**) to (**d**): high curvature, medium curvature, rectilinear and angular

2 Paired-Comparison Task

All the possible six pairs of the four façades (Fig. 1) were presented to all respondents in a random order without repetitions. Stimuli were projected on a screen at a distance of two metres approximately. All participants viewed each pair for 3 s and recorded the preferred façade on a sheet of paper provided individually. The dominance matrix shown in Table 1 summarizes results of paired comparisons. Positive entries represent the number of times the row façade was preferred to the column façade, and main diagonal elements are conventionally set to zero. All the corresponding off-diagonal elements satisfy a constant sum property (i.e., all pairs of corresponding entries (i, j) and (j, i) sum up to 24), resulting in the sum of row and column totals for each façade being also constant. Thanks to this way of representing data, we can easily obtain the façades preference order by the row totals of the dominance matrix, that is—from the most to the less preferred—A (high curvature), B (medium curvature), D (angular), C (rectilinear).

Another consequence of the previous properties is that symmetry is not interesting in this matrix, but it is worthwhile to focalize on the skew-symmetric information. In linear algebra it is known that any square matrix $\mathbf{\Omega} = \{\omega_{ij}\}$ can be additively and uniquely decomposed into a symmetric part $\mathbf{M} = \{m_{ij}\} = \{0.5(\omega_{ij} + \omega_{ji})\}$ and a skew-symmetric part $\mathbf{N} = \{n_{ij}\} = \{0.5(\omega_{ij} - \omega_{ji})\}$, with $\mathbf{\Omega} = \mathbf{M} + \mathbf{N}$. The matrix $\mathbf{M}$ is the best symmetric least squares approximation to $\mathbf{\Omega}$, the matrix $\mathbf{N}$ describes the departures from symmetry and $n_{ij} = -n_{ji}$ holds for each pair (i, j). For the dominance matrix shown in Table 1, all the symmetric entries m_{ij} $(i \neq j)$ are equal to 12, the skew-symmetric entries n_{ij} are the difference of the corresponding frequency in the matrix by the value 12, which in our experiment corresponds to the situation of equilibrium (12 subjects prefer one façade and other 12 subjects prefer the other one).

In his pioneering paper Gower [3] proposed to represent the skew-symmetric component $\mathbf{N}$ by singular value decomposition. The interpretation of the diagram obtained by singular vectors is not in terms of distances but in terms of areas, in particular the area of triangles that pairs of points form with the origin is proportional to the size of skew-symmetry, whose sign is given by the plane orientation. A more detailed description of the non-Euclidean geometry of this type of diagrams can be found in [3]. The skew-symmetric component of the dominance matrix in Table 1 is represented in Fig. 2. The preference order A, B, D, C is easily detected in the diagram going from point A to point C in counter-clockwise direction (skew-symmetry is positive). Façades A and C have the largest imbalance

Table 1 Dominance matrix for the paired-comparison task

Façades	A	B	C	D
A	0	21	22	21
B	3	0	19	17
C	2	5	0	6
D	3	7	18	0

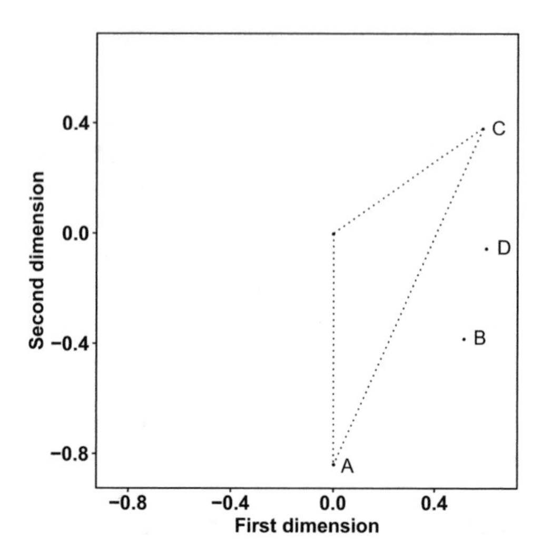

Fig. 2 Gower diagram for the skew-symmetric component of data in Table 1

between each other, so the area of the triangle the two corresponding points form with the origin is large (A dominates C because comes first in counter-clockwise direction, see Fig. 2). Façades B and D have the smallest imbalance so the area of the corresponding triangle is small, B dominates D.

To make easier diagram interpretation, methods based on distance models were also considered to represent skew-symmetry. Bove [2] proposed a method of asymmetric multidimensional scaling that adapted the idea originally proposed by Okada and Imaizumi [6] for asymmetric proximities to skew-symmetric data. The graphical representation is obtained by the following two steps.

Step 1

In the first step, sizes of skew-symmetries, given by the absolute values $|n_{ij}|$, are represented by distances in a *low-dimensional* Euclidean space (usually two-dimensional) according to the following model:

$$f\left(|n_{ij}|\right) = d_{ij} + \varepsilon_{ij} = \sqrt{\sum_{s=1}^{r} \left(x_{is} - x_{js}\right)^2} + \varepsilon_{ij} \tag{1}$$

where f is a chosen data transformation function (e.g., interval, ratio, ordinal transformations); d_{ij} is the distance between façade i and façade j ($d_{ij} = d_{ji}$); x_{is} and x_{js} are the coordinates on dimension s, respectively, of façade i and façade j and ε_{ij} is a residual term. The model can be easily estimated by standard statistical software containing symmetric multidimensional scaling routines. An advantage of this model is that it is easy to incorporate both non-metric approaches and external information regarding the objects.

Step 2

In the second step, the signs of skew-symmetry (and so the preference order) are derived from the comparison of circles represented around the points of the configuration obtained from step 1. The radii r_i of the circles are estimated by the following model:

$$\gamma_{ij} = \left(r_i - r_j\right) + \varepsilon_{ij} \tag{2}$$

where $\gamma_{ij} = 1$ if n_{ij} is positive and $\gamma_{ij} = -1$ if n_{ij} is negative. As a result, when the circle around point i is larger than circle around point j, the estimate of γ_{ij} is positive and the estimate of γ_{ji} is negative (consequently, estimate of skew-symmetry n_{ij} is positive and estimate of skew-symmetry n_{ji} is negative). A least squares solution for the r_i's is

$$\widehat{r_i} = \frac{1}{n}\sum_{j=1}^{n}\gamma_{ij} \tag{3}$$

with $\sum_{i=1}^{n}\widehat{r_i} = 0$, being matrix $\boldsymbol{\Gamma} = \left(\gamma_{ij}\right)$ skew-symmetric. Any translations $\widehat{r_i} + c$ by a constant c is equivalent to the initial solution. However, it is convenient to choose only between solutions with no negative $\widehat{r_i}$'s, because they represent radii. In this application, we chose the unique solution having $min\left(\widehat{r_i}\right) = 0$.

In our application, step 1 was performed by the PROXSCAL program with a transformation ratio option for $|n_{ij}|$, radii in step 2 were computed by a Matlab routine. The method represents the architectural façades as points in a two-dimensional diagram (Stress-I=0,11). Both the façade preference orders and the imbalances are represented: the former as circles with different radii (larger circles correspond to higher ranks of preference), the latter as the distances between points (larger distances correspond to lower equilibrium). The results are shown in Fig. 3.

The overall preference order (A, B, D, C) is represented by the size of the circles. Façade A is the most preferred and is liked equally more than B, C and D. Façades B and D have the smallest imbalance between each other, so they are represented as closer on the plane. Façade C is represented with no ray, so it is dominated by all the other façades, but much more by A and B that are positioned further away from it.

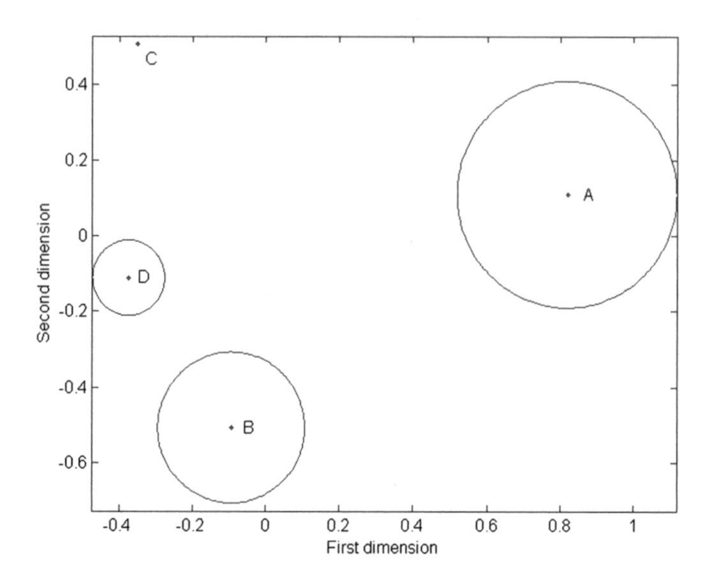

Fig. 3 Asymmetric multidimensional scaling representation for data in Table 1

Table 2 Order choices in the ranking task

	First	Second	Third	Fourth
A	15	5	3	1
B	5	13	6	0
C	1	0	2	21
D	3	6	13	2

3 Ranking Task

In this task, we showed the four façades at the same time on a projection screen, assigning to each of them a corresponding letter, placed at the bottom of the picture. Participants had a grid printed on a sheet of paper. The grid consisted of four boxes with growing numbers, from 1 to 4. People had to write on the corresponding row the façade's letter, according to their preferences. We asked them to classify the façades from the most (= 1) to the least (= 4) preferred. Table 2 reports the number of times each façade (row) was chosen in an order position (column) by participants. The highest frequency in each row of the table allows to confirm the preference order showed in the paired-comparison task: A, B, D, C.

Besides, we analysed the (24 × 4) ranking data matrix $\mathbf{P} = \left(p_{ij}\right)$ with multidimensional unfolding technique (e.g., [1], Chaps. 14–16) to represent relationships between subjects and façades. The unfolding model can be expressed in scalar notation as

$$f\left(p_{ij}\right) = d_{ij}^{unf} + \varepsilon_{ij} = \sqrt{\sum_{s=1}^{r}\left(x_{is} - y_{js}\right)^2} + \varepsilon_{ij} \qquad (4)$$

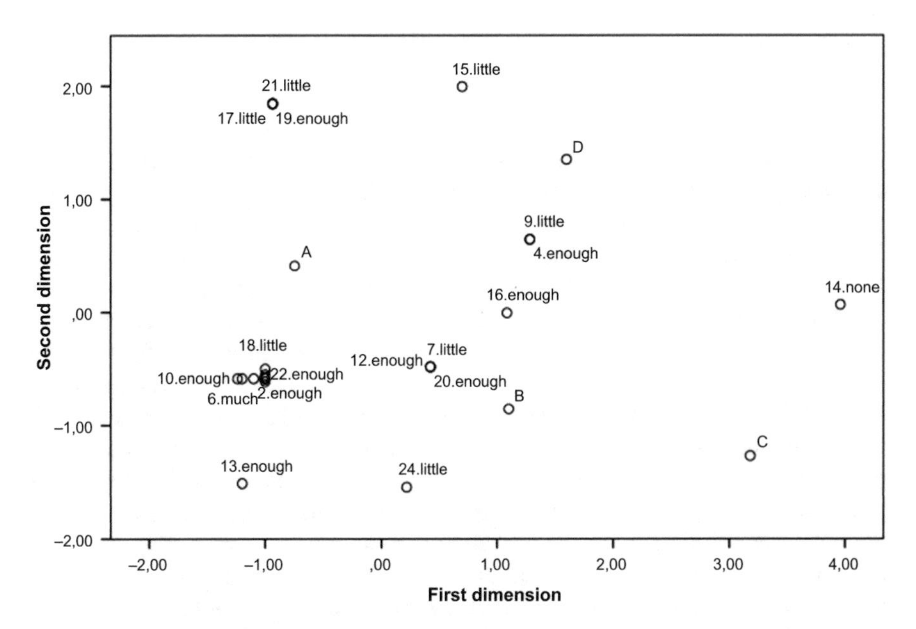

Fig. 4 Multidimensional unfolding representation for rank order scores (subject artistic education level labels: none, little, enough, much)

where, as before, f is a chosen data transformation function; d_{ij}^{unf} is the distance between subject i and façade j (so it can be $d_{ij}^{unf} \neq d_{ji}^{unf}$); x_{is} and y_{js} are the coordinates on dimension s, respectively, of subject i and façade j and ε_{ij} is a residual term. A diagram for the pattern of relationships is obtained by coordinates x_{is} and y_{is}, so that the distances d_{ij}^{unf} from subject points i to façade points j correspond to the rank order scores p_{ij}, with high rank order scores corresponding to small distances. Moreover, this method can be easily applied by standard statistical software containing multidimensional unfolding routines. The results obtained with PREFSCAL program are shown in Fig. 4, where numbers represent the subjects and letters represent the façades (Stress-I=0,07). Labels attached to each subject number represent artistic education levels. According to the unfolding model properties, the subjects tend to be closer to the façades for which they expressed a higher rank in the task. Overall, façade A—high curvature—and to a less extent façade B—medium curvature—are the two stimuli around which the majority of the subjects is placed, being also the one with the higher artistic education. Subjects 4, 9 and 15 preferred façade D, but their artistic education is positioned at a medium-low level. Only one subject (subject 14) preferred façade C, but she has the lowest level of artistic education, corresponding to no artistic education at all.

4 Conclusions

In this article, we showed that graphical representations obtained by multidimensional scaling and unfolding allow to easily detect preference order, size of asymmetry and relationships between subjects and stimuli. In our experiments it was confirmed that curvature influences preferences also for architectural stimuli. The high curvature façade was the most preferred, followed by the medium, angular and rectilinear ones. The rectilinear stimulus was always the least preferred and not the sharp one as expected. This result provides an important insight to better understand human preferences, suggesting that the curvature effect can be modulated by controlling for the level of sharpness of the stimuli it is compared with. In line with previous research showing that expertise plays an important role in influencing aesthetic judgments for sharp stimuli, our study showed that participants with relatively poor art training preferred rectilinear façades, while people with higher levels of artistic training preferred the curved ones. Due to the non-probabilistic features of our small sample, we followed a 'data-analytic' approach emphasizing the graphical display of data. Future developments of this research will consider selection of large probabilistic samples in order to confirm our hypothesis and generalize our results in a more formalized context (e.g., [5, 8]).

Acknowledgements We thank the architect Stefania Lamaddalena for sharing with us her professional knowledge and for producing the AutoCAD 2-D render of the stimuli we used in this study.

References

1. Borg, I., Groenen, P.J.F. : Modern multidimensional scaling. Theory and Applications, 2nd edn. Springer, New York (2005)
2. Bove, G.: Exploratory approaches to seriation by asymmetric multidimensional scaling. Behaviormetrika **39**, 63–73 (2012)
3. Gower, J.C.: The analysis of asymmetry and orthogonality. In: Barra, J.R., et al. (eds.) Recent Developments in Statistics, pp. 109–123. North Holland, Amsterdam (1977)
4. Mastandrea, S., Bartoli, G., Carrus, G.: The automatic aesthetic evaluation of different art and architectural styles. Psychol. Aesthet. Creat. Arts **5**(2), 126–134 (2011)
5. Maydeu-Olivares, A., Bockenholt, U.: Modeling preference data. In: Millsap, R.E., Maydeu-Olivares, A. (eds.) The Sage Handbook of Quantitative Methods in Psychology, pp. 264–282. Sage, Los Angeles (2009)
6. Okada, A., Imaizumi, T.: Nonmetric multidimensional scaling of asymmetric proximities. Behaviormetrika **21**, 81–96 (1987)
7. Palumbo, L., Ruta, N., Bertamini, M.: Comparing angular and curved shapes in terms of implicit associations and approach/avoidance responses. PLoS One, **10**(10), e0140043 (2015)
8. Piccolo, D.: Observed information matrix for MUB models. Quad. Stat. **8**, 33–78 (2006)

Community Structure in Co-authorship Networks: The Case of Italian Statisticians

Domenico De Stefano, Maria Prosperina Vitale, and Susanna Zaccarin

Abstract Community detection is a very appealing topic in network analysis. A precise definition of community is still lacking, so the comparison of different methods is not a simple task. This paper shows exploratory results by adopting two well-known community detection methods and a new proposal to discover groups of scientists in the co-authorship network of Italian academic statisticians.

Keywords Co-authorship networks · Community detection algorithms · Modularity · Italian statisticians

1 Introduction

In the last decades social network analysis (SNA) has become a widespread methodological approach to study scientific collaboration. As stated in several studies [5, 8], scientific collaboration is a crucial factor to enhance publication productivity and research quality. The role of scientific collaboration allowing a fertile ground for the development of new ideas is also recognized in research funding European programmes as well as national projects.

Thanks to the availability of international bibliographic archives, co-authorship networks—in which the connection between two researchers is given by the number

D. De Stefano
Department of Social and Political Sciences, University of Trieste, Trieste, Italy
e-mail: ddestefano@units.it

M. P. Vitale
Department of Political and Social Studies, University of Salerno, Fisciano, Italy
e-mail: mvitale@unisa.it

S. Zaccarin (✉)
Department of Business, Economic, Mathematics and Statistics, University of Trieste, Trieste, Italy
e-mail: susanna.zaccarin@deams.units.it

F. Greselin et al. (eds.), *Statistical Learning of Complex Data*,
Studies in Classification, Data Analysis, and Knowledge Organization,
https://doi.org/10.1007/978-3-030-21140-0_7

of papers they co-authored—are used as a proxy of scholars' collaborative behavior in science [2]. Usually, binary networks—setting the connections greater than zero to one—are considered in empirical analysis. A common aim in co-authorship studies through SNA perspective is the understanding of network properties since the evolution of topics and methods in scientific fields appears strongly related to the topological structure of the collaboration patterns among scholars. In this stream of research, the recovery of *communities*—the term used to identify groups or clusters of actors in a graph—shaping the network structure sounds very appealing and informative. Unfortunately, a precise definition of what constitutes a community—broadly, part of a network where internal links are denser than external ones—is still lacking [16]. As a consequence of this conceptual vagueness, several community detection algorithms have been proposed in the literature [9].

Starting from previous findings on small-world topology in the co-authorship network of Italian academic statisticians [6, 11], the present contribution intends to deepen the analysis of this case study uncovering a meaningful community structure for Italian scholars. To this aim, results from three community detection methods, the Girvan–Newman algorithm [13], the Louvain algorithm [3], and a new method—modal clustering algorithm [12]—will be compared. The evaluation of performance measures [16] and the interpretation of main results should benefit from the common clustering perspective shared by the three algorithms.

This paper is organized as follows: Section 2 reviews the main characteristics of the three methods and their performance in identifying communities within an illustrative example. Section 3 discusses the main results obtained by adopting the aforementioned methods on the co-authorship network of Italian statisticians using also available scholar's attributes (i.e., scientific field and university affiliation). Section 4 reports new lines of research for future work.

2 Community Detection Methods

Similarly to the problem of clustering for attribute data, the lack of a unique definition of community in the presence of network data has led to the proliferation of several methods in different theoretical contexts. Among them, some are explicitly designed to handle these kinds of data. For instance, blockmodeling [7, pp. 11–12] is a methodological approach "*to identify, in a given network, clusters of actors that share structural characteristics in terms of some relations*," mainly based on partitioning the relational matrix into a set of blocks.

Recently, a huge variety of network-based clustering techniques, the so-called community detection methods, have been developed based on hierarchical clustering techniques [13], locating network communities by statistical analysis of the raw data [14], or optimizing different quality functions [9].

In the following, we focus on two well-known community detection algorithms, and a new method based on an adaptation to network data of modal clustering procedure (for an overview with standard data, see [1]):

1. the Girvan–Newman algorithm [13], one of the most popular community detection approach. It is based on a hierarchical divisive procedure in which links are iteratively removed based on the value of the edge's betweenness. The procedure of link removal ends when the value of the modularity index Q is maximized. This index [4, 13] measures the fraction of the edges in the network that connect nodes within-community minus its expected value in the case of a network with edges placed at random. It assumes a minimum value of 0, when the number of within-community edges is no better than the randomized network, and a maximum value of 1 in the presence of strong community structure. The index usually falls in the range 0.3–0.7, and a value of around 0.3 is a good indicator of significant community structure in the network;
2. the Louvain algorithm [3], also based on the modularity index and on a hierarchical approach. Initially, each node is assigned to a community on its own. In every step, nodes are re-assigned to communities in a local, greedy way: each node is moved to the community in which it achieves the highest contribution to the modularity;
3. the modal clustering algorithm [12], which starts from the idea that highly connected sets of nodes can be detected around the modes of a "density" function f reflecting the cohesiveness between nodes—e.g., centrality measures [10] like the node degree (i.e., the number of links a node has with the other nodes in the network) or the actor betweenness (i.e., the number of those shortest paths passing through a specific node connecting two other nodes). The modes of f are seen as the archetypes of the clusters, which are in turn represented by their surrounding regions. Any section of f, at a level λ, identifies a level set, namely the region with f value above λ. The key idea is that when f is unimodal, there is no clustering structure, and the level set is connected for any choice of λ. Conversely, when f is multimodal, the identified level set may be connected or not, depending on λ value. In particular, nodes are clustered together when they have a value of f above the examined threshold λ and they are connected in the underlying network. Clustering is performed around the modal actors, namely actors showing the largest value of the chosen function. Furthermore, by varying the level set the method gives rise to a tree diagram, called cluster tree (which is graphically similar to a dendrogram), where each leaf corresponds to a mode of the function.

The first two algorithms are particularly suited for undirected and unweighted relational data (likewise the most usual case of co-authorship data obtained disregarding the number of papers co-authored by pairs of scholars), while the third one is more flexible since different concepts of cohesiveness among actors can be used.

To compare the three approaches in discovering communities, we consider the Zachary's karate club network data [17] describing the friendship relationship among 34 members of a karate club at a US university in the 1970s. A useful

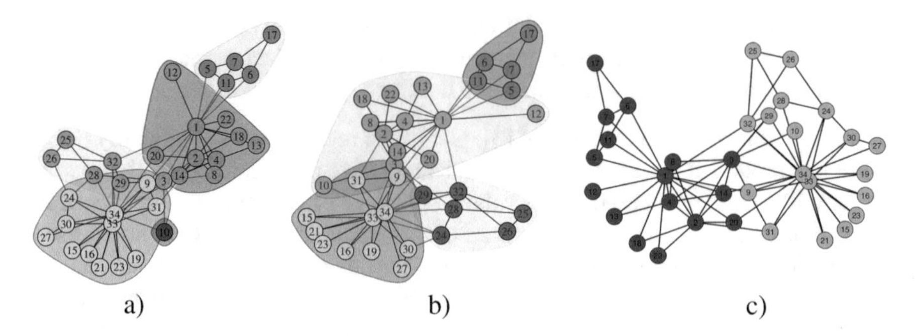

Fig. 1 Comparison of the three community detection methods for Zachary's karate club network data: (**a**) Girvan–Newman algorithm; (**b**) Louvain algorithm; (**c**) modal clustering algorithm

feature of this dataset is that, during the period of observation, the club split into two factions, due to a dispute between the administrator and the karate instructor. Thus, a true cluster membership of the actors in the network is known and can be used as a benchmark to evaluate the performance of different methods. Figure 1 shows the communities identified by using the three algorithms. It is possible to appreciate that the modal clustering method, using node degree as density function to reflect actors' cohesiveness, allows to detect the two factions underlying the networks. In particular, the method works by clustering every actors around the modal actors—that are the two most central ones in terms of their degree in Fig. 1c—that, incidentally, are the members around which the karate club splits into two distinct factions. The other approaches are able to detect different partitions, in particular consisting of four groups.

3 Community Detection Results for Italian Statisticians

The three aforementioned community detection methods are used to analyze the co-authorship network defined for the population of the 792 Italian academic statisticians belonging to five scientific subfields,[1] as recorded in the Italian Ministry of University and Research (MIUR) database at March 2010. To collect publications three bibliographic archives—two international (Web of Science and Current Index to Statistics) and one national based on publications attached to the nationally funded grants (PRIN projects)—are considered [6]. Hence the co-authorship network under analysis is the result of combining multiple data sources [11].

[1]The five subfields established by the Italian governmental official classification are: Methodological Statistics, Statistics for Experimental and Technological Research, Economic Statistics, Demography, and Social Statistics.

The general aim of the community detection procedures here adopted is to discover if the co-authorship network of Italian statisticians can be clustered into communities. In order to let the results comparable, the three community detection methods are performed on the largest connected component of the given graph (i.e., giant component). This approach, recognized in the related literature in order to isolate disjoint components [15], is useful in our case given that only the modal clustering algorithm is able to handle disconnected graphs. In the observed co-authorship network, the giant component consists of 660 authors, representing the 82% of statisticians. Therefore the analysis can be restricted to this set of authors without loss of generality. In performing the modal clustering method, two different density functions (degree and betweenness) are chosen.

The main results of the three procedures are reported in Table 1. In general, the methods are quite comparable in terms of number of detected communities and of their sizes. The Girvan–Newman algorithm produces the larger number of communities (#. 22). Also the quality of the partitions, measured by the modularity index Q, is quite similar across methods. The lower value is associated with modal clustering with the betweenness as density function that is the method that also gives raise to communities of relative larger sizes with respect to the other two methods.

The modal clustering (with degree as density function) and the Louvain algorithm show the highest—and similar—values of the modularity index as well as the same total number of detected communities (#. 18). In the following, the composition of the first 9 larger communities identified by these two approaches is analyzed. These larger communities are quite representative since for both methods they comprise about the 70% of the 660 statisticians in the giant component.

Table 2 reports some descriptive measures of the 9 communities listed in descending order by size. In both algorithms, the detected communities share quite similar structural characteristics. By way of example, the largest community (C1) comprises 91 and 69 statisticians, for the modal clustering and the Louvain algorithm, respectively.

The author average degree—computed within the community—is usually comparable across methods, ranging from a minimum of 1.75 (community C4 by modal clustering) to a maximum of 4.04 (community C3 for Louvain algorithm). The ratio between within-community links (edges representing the relationship in the same community) and the external links (edges activated with nonmembers of the

Table 1 Performance measures of giant component of the Italian statisticians co-authorship network by methods

Method	C	Average (St. Dev.)	Q
Girvan–Newman	22	30.000 (15.754)	0.752
Louvain	18	36.667 (17.283)	0.762
Modal clustering (betweenness)	13	50.769 (30.444)	0.702
Modal clustering (degree)	18	36.667 (23.118)	0.761

C = #. of detected communities, Average = Average number of authors in communities (St. Dev.), Q = modularity index

Table 2 Descriptive measures of the first nine detected communities obtained by the modal clustering (MC) and the Louvain algorithms for the giant component of the Italian statisticians co-authorship network

	Size		Average degree author		Intra-extra links ratio	
Community	MC (degree)	Louvain	MC (degree)	Louvain	MC (degree)	Louvain
C1	91	69	3.52	2.92	0.120	0.056
C2	67	65	2.48	3.75	0.063	0.092
C3	57	53	2.60	4.04	0.057	0.022
C4	49	49	1.75	4.00	0.011	0.021
C5	49	49	2.00	2.98	0.016	0.021
C6	48	48	2.00	3.29	0.039	0.026
C7	48	44	2.17	3.36	0.018	0.076
C8	47	41	3.23	3.61	0.038	0.056
C9	44	40	2.32	3.35	0.036	0.050

community) is quite small for both methods. Looking at the internal composition by scientific subfield and university affiliation, in the Louvain method, the largest community includes several authors in the statistics subfield.

In the modal clustering, the emerging largest community is composed mostly of authors in statistics subfield and some authors in economic statistics subfield, mainly clustered according to the geographic proximity of their universities. In particular, the majority of authors in this cluster are affiliated to the universities located in the North and in the Center of Italy (e.g., Florence, Padua, Rome, and Milan). The same differences arise looking at the composition of the other larger detected communities. Both methods find clusters that are homogeneous by scientific sectors (demographers and social statisticians, on the one hand, and methodological statisticians, on the other hand, tend to create strong communities), although it seems that modal clustering groups together authors on the basis of links mainly driven by the geographic proximity of the universities in which they are affiliated, while Louvain algorithm aggregates authors on the basis of network characteristics.

Generally speaking, comparing all possible couples of communities, the overlapping among the detected communities is low. The average Jaccard index is indeed equal to 0.02. Only some communities present a sort of overlapping with about 30% of common members, as showed in the example in Fig. 2 for community 1 (C1) in the Louvain algorithm and community 4 (C4) in the modal clustering algorithm. These methods are therefore able to capture common relational aspects of the observed co-authorship network enriching the interpretation of the findings related to the authors' attributes.

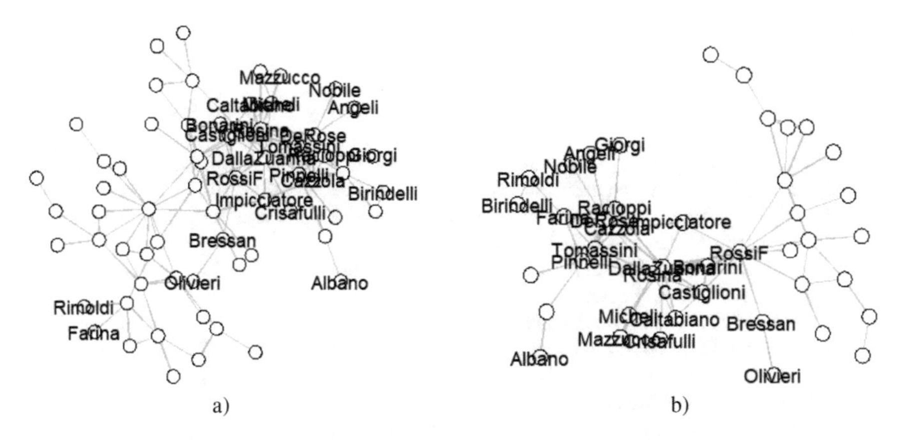

Fig. 2 Representation of the communities with the largest overlapping number of actors: (**a**) community 1 (C1) Louvain algorithm; (**b**) community 4 (C4) modal clustering algorithm. The names of the statisticians common to both communities are displayed

4 Conclusions

The general aim of the community detection procedures here adopted was to discover if the co-authorship network of Italian statisticians can be clustered into communities. To this purpose, results from three different community detection methods, the Girvan–Newman algorithm, the Louvain algorithm, and the modal clustering algorithm, have been compared by presenting performance measures and specific internal communities interpretations. The most suitable methods in terms of quality of the partitions discovered are the modal clustering algorithm and the Louvain algorithm.

As general evidence, it seems that the co-authorship network of the Italian statisticians is clustered in a relatively small number of communities with different internal composition that is mainly determined by authors' scientific field and university affiliation.

In order to find denser communities it would be important to consider in the analysis also the strength of the collaboration relationship by using the number of co-authored papers among couples of authors. As future line of research we will intend to extend the described community detection methods to weighted networks. It also would be interesting to explore the community structures dealing with the presence of multiplex networks, when collaboration is described by measuring also other kinds of relationships among scientists (e.g., co-participation on funded projects).

References

1. Azzalini, A., Torelli, N.: Clustering via nonparametric density estimation. Stat. Comput. **17**(1), 71–80 (2007)
2. Bellotti, E., Kronegger, L., Guadalupi, L.: The evolution of research collaboration within and across disciplines in Italian academia. Scientometrics **109**, 783–811 (2016)
3. Blondel, V.D., Guillaume, J.-L., Lambiotte, R., Lefebvre, E.: Fast unfolding of communities in large networks. J. Stat. Mech. **2008**(10), P10008 (2008)
4. Clauset, A., Newman, M.E.J., Moore, C.: Finding community structure in very large networks. Phys. Rev. E **70**, 066111 (2004)
5. De Stefano, D., Zaccarin, S.: Co-authorship networks and scientific performance: an empirical analysis using the generalized extreme value distribution. J. Appl. Stat. **43**, 262–279 (2016)
6. De Stefano, D., Fuccella, V., Vitale, M.P., Zaccarin, S.: The use of different data sources in the analysis of co-authorship networks and scientific performance. Soc. Netw. **35**, 370–381 (2013)
7. Doreian, P., Batagelj, V., Ferligoj, A.: Generalized Blockmodeling. Structural Analysis in the Social Sciences. Cambridge University Press, New York (2005)
8. Ferligoj, A., Kronegger, L., Mali, F., Snijders, T.A.B., Doreian, P.: Scientific collaboration dynamics in a national scientific system. Scientometrics **104**, 985–1012 (2015)
9. Fortunato, S., Hric, D.: Community detection in networks: a user guide. Phys. Rep. **659**, 1–44 (2016)
10. Freeman, L.C.: Centrality in social networks conceptual clarification. Soc. Netw. **1**, 215–239 (1978)
11. Fuccella, V., De Stefano, D., Vitale, M.P., Zaccarin, S.: Improving co-authorship network structures by combining multiple data sources: evidence from Italian academic statisticians. Scientometrics **107**, 167–184 (2016)
12. Menardi, G., De Stefano, D.: Modal clustering of social network. In: Cabras, S., Di Battista, T., Racugno, W. (eds.) Proceedings of the 47th SIS Scientific Meeting of the Italian Statistical Society. CUEC Editrice, Cagliari (2014)
13. Newman, M.E.J., Girvan, M.: Finding and evaluating community structure in networks. Phys. Rev. E **69**, 026113 (2004)
14. Palla, G., Derényi, I., Farkas, I., Vicsek, T.: Uncovering the overlapping community structure of complex networks in nature and society. Nature **435**, 814–818 (2005)
15. Savić, M., Ivanović, M., Radovanović, M., Ognjanović, Z., Pejović, A., Krüger, T.J.: Exploratory analysis of communities in co-authorship networks: a case study. In: ICT Innovations 2014, pp. 55–64. Springer, Berlin (2015)
16. Schaub, M.T., Delvenne, J.C., Rosvall, M., Lambiotte, R.: The many facets of community detection in complex networks. Appl. Netw. Sci. **2**(1), 4 (2017)
17. Zachary, W.W.: An information flow model for conflict and fission in small groups. J. Anthropol. Res. **33**(4), 452–473 (1977)

Analyzing Consumers' Behavior in Brand Switching

Akinori Okada and Hiroyuki Tsurumi

Abstract Asymmetric multidimensional scaling is extended to represent differences among consumers in brand switching. The asymmetric multidimensional scaling, based on the singular value decomposition, represents asymmetric relationships among brands in the brand switching by introducing the outward tendency which corresponds to the left singular vector and the inward tendency which corresponds to the right singular vector. The resulting configuration is represented in a plane spanned by the left and the right singular vectors where each brand is represented as a point. Each dimension (component) has its own plane or a two-dimensional configuration. The asymmetric multidimensional scaling is extended so that each consumer is represented as a point in the plane. The joint configuration of brands and consumers represents how each consumer or a group of consumers relates to brands in the brand switching. The procedure is applied successfully to brand switching data among potato snacks.

Keywords Asymmetry · Brand switching · Consumer · Individual differences · Multidimensional scaling

1 Introduction

The brand switching is derived by comparing brands purchased in two consecutive periods by a consumer. Asymmetric multidimensional scaling has been used to analyze brand switching [5, 6]. While these studies represent asymmetric relation-

A. Okada (✉)
Research Institute, Tama University, Tokyo, Japan
e-mail: okada@rikkiyo.ac.jp

H. Tsurumi
Yokohama National University, Hodogaya-ku, Yokohama-shi, Japan
e-mail: tsurumi@ynu.ac.jp

F. Greselin et al. (eds.), *Statistical Learning of Complex Data*,
Studies in Classification, Data Analysis, and Knowledge Organization,
https://doi.org/10.1007/978-3-030-21140-0_8

ships of brand switching, they cannot analyze the differences among consumers nor their relationships with brands in the brand switching. It is important to know differences among consumers [3] and to disclose relationships between consumers and the brand switching they did [1]. The present study extends the asymmetric multidimensional scaling so that differences among consumers can be represented. In particular, our approach allows also to visualize how each consumer or a group of consumers relates to brands in the brand switching.

2 Method

The asymmetric multidimensional scaling based on the singular value decomposition [2] is briefly described below. Let $\mathbf{A}$ be an $n \times n$ matrix of asymmetric brand switching matrix, where n is the number of brands. The (j, k) element of $\mathbf{A}$ represents the frequency of the brand switching from brands j to k, where brand j is purchased at the first period, and brand k is purchased at the second period. By the singular value decomposition, $\mathbf{A}$ is approximated by using r dimensions (components) as;

$$\mathbf{A} \simeq \mathbf{XDY}',$$

where $\mathbf{D}$ is the $r \times r$ diagonal matrix of r largest singular values $(d_1, \ldots, d_r)$ in descending order at its diagonal elements, $\mathbf{X}$ is the $n \times r$ matrix of corresponding left singular vectors (the length is unity), and $\mathbf{Y}$ is the $n \times r$ matrix of corresponding right singular vectors (the length is unity). The jth element of the ith column of $\mathbf{X}$ represents the outward tendency of brand j along Dimension i, and the kth element of the ith column of $\mathbf{Y}$ represents the inward tendency of brand k along Dimension i.

Let $\mathbf{S}$ be an $N \times n$ matrix where each row has one element of 1 and $(n-1)$ elements of 0, where N is the number of consumers. If the (m, j) element of $\mathbf{S}$ is 1, consumer m purchased brand j at the first period. Similarly, $\mathbf{T}$ is an $N \times n$ matrix where each row has one 1 and $(n-1)$ elements are 0. If the (m, k) element of $\mathbf{T}$ is 1, consumer m purchased brand k at the second period. $\mathbf{A}$ can be derived by

$$\mathbf{A} = \mathbf{S}'\mathbf{T}.$$

Thus $\mathbf{S}'\mathbf{T} = \mathbf{A} \simeq \mathbf{XDY}'$. We define $\mathbf{F} = \mathbf{SX}$ and $\mathbf{G} = \mathbf{TY}$, where $\mathbf{F}$ and $\mathbf{G}$ are the $N \times r$ matrices. The mth row of $\mathbf{F}$ represents the outward tendency of consumer m along r dimensions. The mth row of $\mathbf{G}$ represents the inward tendency of consumer m along r dimensions. By deriving the outward and inward tendencies of a consumer or a group of consumers in the planar configuration of brands, relationships of a consumer or a group of consumers with brands in the brand switching are shown.

Table 1 Brand switching matrix among nine potato snack brands

	Period 2								
Period 1	A	B	C	D	E	F	G	O	Z
Brand A	140	15	8	3	7	0	3	15	5
Brand B	15	129	12	8	16	0	12	24	12
Brand C	5	16	27	4	9	0	5	91	9
Brand D	2	7	9	32	5	0	1	4	3
Brand E	8	19	5	5	45	0	7	17	14
Brand F	2	9	4	2	1	0	0	7	1
Brand G (component)	1	2	1	2	1	0	9	3	3
Brand O	5	10	7	6	4	0	3	7	4
Brand Z	10	10	6	9	21	0	4	15	22

3 A Real Dataset: Potato Snack Brands and Their Customers

The brand switching data were collected at two periods (period 1: June 2–July 31; period 2: August 1–August 31 of 2009). The frequency of the brand switching in Table 1 is the number of consumers who changed the largest purchase brand from periods 1 to 2. The brand switching matrix is derived from the purchase record of 882 customers who purchased potato snacks at both periods 1 and 2. Table 1 shows a brand switching matrix among nine potato snack brands (A, B, C, D, E, F, G, O, Z). Brand O represents brands other than A, ..., G, and Z. The elements of the sixth column of Table 1 are null, because brand F was withdrawn at period 2. The (j, k) element of Table 1 is the number of consumers whose largest purchase brand was j at period 1 and k at period 2. A further detail is given in [6].

4 Data Analysis and Obtained Results

The 9×9 brand switching matrix shown in Table 1 is asymmetric, and was analyzed by the asymmetric multidimensional scaling. The five largest singular values are 162.1, 122.1, 54.5, 32.6, and 22.5. The three-dimensional result was chosen as the solution same to the earlier study [6]. Each dimension (component) has its own planar configuration: Dimension i has a planar configuration spanned by the left singular vector (abscissa) and the right singular vector (ordinate) correspond to the ith largest singular value (see Figs. 1, 2, and 3). The abscissa represents the outward tendency which tells the weakness of a brand or the easiness to be switched from the brand to the other brands, and the ordinate represents the inward tendency which tells the strength of a brand or the easiness to be switched to the brand from the other brands. The outward and inward tendencies of each consumer are derived by $\mathbf{F} = \mathbf{SX}$ and $\mathbf{G} = \mathbf{TY}$. Then 882 consumers were classified into two groups according to the amount of money they spent on potato snacks, Group 1 consists of

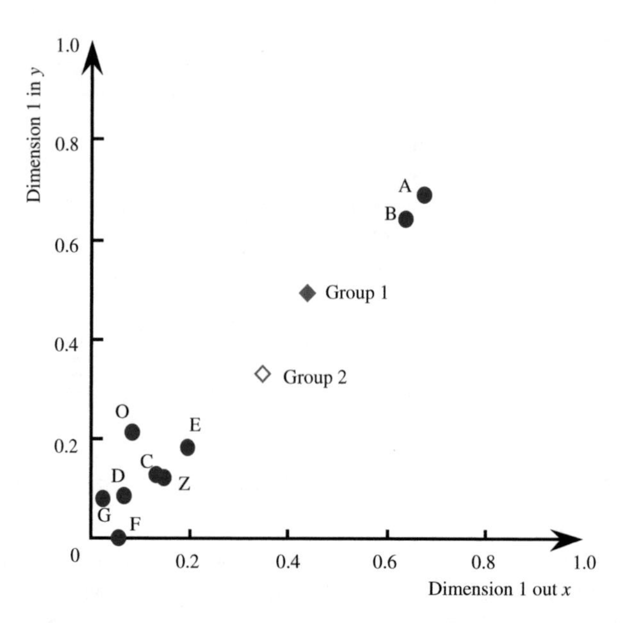

Fig. 1 Joint configurations of brands and groups along Dimension 1

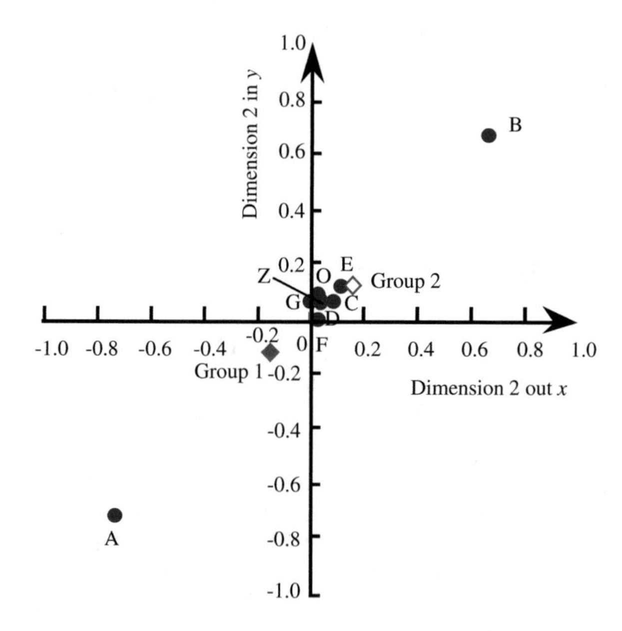

Fig. 2 Joint configurations of brands and groups along Dimension 2

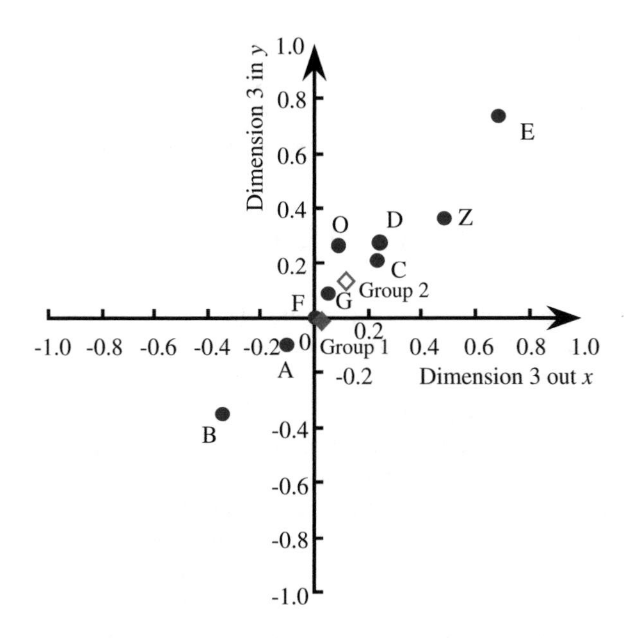

Fig. 3 Joint configurations of brands and groups along Dimension 3

Table 2 Mean outward and inward tendencies for Groups 1 and 2 of consumers: Group 1 (larger and equal to the average) and Group 2 (smaller than the average)

	Dimension 1		Dimension 2		Dimension 3	
Group	Outward	Inward	Outward	Inward	Outward	Inward
1	0.477	0.493	−0.158	−0.120	0.026	−0.013
2	0.353	0.333	0.148	0.130	0.122	0.136

270 consumers whose amounts of money are larger or equal to the average amount of money, and Group 2 consists of 612 consumers whose amounts of money are smaller than the average. The means of outward tendency and inward tendency for Groups 1 and 2 along Dimensions 1, 2, and 3 are shown in Table 2.

The mean outward tendency for a group represents the mean of outward tendencies of brands from which consumers in the group switched to the other brands along each dimension. And the mean inward tendency for a group represents the mean of inward tendencies of brands to which consumers in the group switched from the other brands along each dimension. Two groups can be represented, respectively, as a point in a configuration of brands along each dimension. Along Dimension 1, Group 1 has the larger mean outward tendency than Group 2 has, and this is also true for the mean inward tendency. Along Dimension 1, the mean outward tendency is smaller than the mean inward tendency for Group 1, while the mean outward tendency is larger than the mean inward tendency for Group 2. Along Dimension 2, Group 1 has negative mean outward and inward tendencies, suggesting that the point representing Group 1 is in the third quadrant (Q3) in the configuration along Dimension 2. The point representing Group 2 is in the first quadrant (Q1) in the configuration, because Group 2 has positive mean outward and inward tendencies. Similarly, along Dimension 3, the point representing Group 1 is in Q4, and the point representing Group 2 is in Q1 in the configuration.

5 Discussion of the Obtained Findings

A joint configuration of brands and groups tells relationships among brands, among groups, and between brands and groups as well. Figure 1 shows the joint configuration of brands and two groups along Dimension 1. Group 1 is closer to brands A and B than Group 2 is, and Group 2 is closer to brands other than A and B than Group 1 is. Brands from which Group 1 switched to the other brands have the larger outward tendency than brands from which Group 2 switched to the other brands have. This is also true for the inward tendency. These suggest that the brand switching of Group 1 includes larger proportion of brands A and B (which have larger outward and inward tendencies and are nearer to Group 1 than to Group 2 in the configuration) than that of Group 2 includes, while the brand switching of Group 2 includes larger proportion of brands other than A and B than Group 1 includes (which have the smaller outward and inward tendencies and are nearer to Group 2 than to Group 1 in the configuration). This is validated by the figures in Table 3.

Table 3 shows the number of brand switchings from/to brand A or B and other brands for each of Groups 1 and 2. For the brand switching of Group 1, 171 (171/270=0.63) of 270 were done from brand A or B, and 99 (99/270=0.37) were done from brands other than A and B. For the brand switching of Group 2, 253 (253/612=0.41) of 612 were done from brand A or B, and 359 (0.59) were done from brands other than A and B. For the brand switching of Group 1, 180 (0.67) brand switchings were done to brand A or B, and 90 (0.33) were done to brands other than A and B. For the brand switching of Group 2, 225 (0.37) brand switchings were done to brand A or B, and 387 (0.63) brand switchings were done to brands other than A and B. These figures show that the brand switching of Group 1 more closely relates to brands A and B, and less closely relates to brands other than A and B, and that the brand switching of Group 2 less closely relates to brands A and B, and more closely relates to brands other than A and B.

Figure 2 shows the joint configuration of brands and groups along Dimension 2. Only brand A is in Q3. The other brands are in Q1. Group 1 is in Q3. This suggests

Table 3 Brand switching from/to brand A or B and other brands of Groups 1 and 2 of consumers

Brand switching	Group 1 (270)	Group 2 (612)
From A or B	171 (0.63)	253 (0.41)
From other than A or B	99 (0.37)	359 (0.59)
To A or B	180 (0.67)	225 (0.37)
To other than A or B	90 (0.33)	387 (0.63)

Table 4 Brand switching from/to brand A in the third quadrant (Q3) and brands in the first quadrant (Q1) of Groups 1 and 2 of consumers

Brand switching	Group 1 (270)	Group 2 (612)
From A	119 (0.44)	77 (0.13)
From brands in Q1	151 (0.56)	535 (0.87)
To A	115 (0.43)	73 (0.12)
To brands in Q1	155 (0.57)	539 (0.88)

that the brand switching of Group 1 includes brand A more than that of Group 2 does. Group 2 is in Q1, suggesting that the brand switching of Group 2 includes brands in Q1 more than that of Group 1 does.

Table 4 shows the number of brand switchings from/to brand A (in Q3) and brands in Q1 for each of Groups 1 and 2. The brand switching from brand A is 119 (0.44) for Group 1, while the brand switching from brand A is 77 (0.13) for Group 2. The brand switching to brand A is 115 (0.43) for Group 1, while the brand switching to brand A is 73 (0.12) for Group 2. The brand switching from brands in Q1 is 151 (0.56) for Group 1, while the brand switching from brands in Q1 is 535 (0.87) for Group 2. The brand switching to brands in Q1 is 155 (0.57) for Group 1, while the brand switching to brands in Q1 is 539 (0.88) for Group 2. These figures support the suggestion mentioned above.

Figure 3 shows the joint configuration of brands and groups along Dimension 3. Group 1 is in Q4, and is close to the origin. Brands A and B are in Q3, and the other brands are in Q1. In the brand switching, brands in Q4 are dominated by brands in Q3, and are dominant over brands in Q1 [4]. This suggests that the brand switching of Group 1 from brands in Q3, and that to brands in Q1 is larger than the other way round. Group 2 is in Q1, suggesting that the brand switching of Group 2 is closely related to brands in Q1. The number of brand switchings from/to Q1and Q3 is already shown in Table 3, because brands A and B are in Q3, and the other brands are in Q1.

The brand switching of Group 1 from brands in Q3 (brand A or B) is 171 (0.63) and those from brands in Q1 (brands other than A and B) is 99 (0.37), this supports the suggestion mentioned above. But the brand switching to brands in Q3 is 180 (0.67) and that to brands in Q1 is 90 (0.33), this does not support the suggestion. For Group 2, the brand switching from brands in Q1 is 359 (0.59) and that to brands in Q1 is 387 (0.63). This shows that the brand switching of Group 2 is closely related to brands in Q1.

The asymmetric multidimensional scaling based on the singular value decomposition was extended so that a joint configuration of brands and consumers (groups of consumers) is represented in a planner configuration. The joint configuration can represent relationships between consumers (groups of consumers) and brands in the brand switching. The extended asymmetric multidimensional scaling was applied to the brand switching among potato snack brands successfully.

Each dimension represents a different aspect of the relationship between groups of consumers and brands in the brand switching. Dimension 1 represents that the brand switching of Group 1 is closely related to brands A and B, and that the brand switching of Group 2 is closely related to brands other than A and B. This seems reasonable, because brands A and B have the two largest market shares among nine brands [26.4% and 20.5%, 6, Table 1], and Group 1 consists of consumers whose amount of money of purchasing potato snacks is larger or equal than the average. Dimension 2 represents another aspect of the relationship of two groups with brands A and B. It is disclosed that the brand switching of Group 1 is mainly related to brand A, and that the brand switching of Group 2 is substantially related to brand B, while the brand switching between A and B is not large [4]. Dimension 3 represents that the substantial brand switching of Group 1 was done from brand A or B to the other brands, and that the brand switching of Group 2 is mainly related to brands other than A and B. These findings were obtained by the joint configuration of brands and groups of consumers introduced by the adopted methodology of asymmetric multidimensional scaling. In the present study, groups were made up based on the amount of money of purchase. Studies using groups based on characteristics or buying behaviors of consumers seem interesting to be further considered in view of investigating the relationships within consumers and brands in the brand switching.

Acknowledgements The authors would like to express their gratitude to the participants of the ClaDAG2017 Conference at the University of Milano-Bicocca for their comments and remarks. They are deeply grateful to reviewers and editors for their review and comments.

References

1. Colombo, R.A., Morrison, D.G.: A brand switching model with implications for marketing strategies. Market. Sci. **8**, 89–99 (1989)
2. Eckart, C., Young, G.: The approximation of one matrix by another of lower rank. Psychometrika **1**, 211–218 (1936)
3. Grover, R., Srinivasan, V.C.: A simultaneous approach to market segmentation and market structuring. J. Market. Res. **24**, 139–1534 (1987)
4. Okada, A., Hayashii, T.: Asymmetric multidimensional scaling of subjective similarities among occupational categories. In: van der Ark, L.A., Wiberg, M., Culpepper, S.A., Douglas, J.A., Wang, W.C. (eds.) Quantitative Psychology, pp. 129–139. Springer Nature, Cham (2017)
5. Okada, A., Tsurumi, H.: External analysis of asymmetric multidimensional scaling based on singular value decomposition. In: Giudici, P., Ingrassia, S., Vichi, M. (eds.) Statistical Models for Data Analysis, pp. 269–278. Springer, Heidelberg (2013)
6. Okada, A. Tsurumi, H.: Evaluating the effect of new brand by asymmetric multidimensional scaling. In: Vicari, D., Okada, A., Ragozini, G., Weihs, C. (eds.) Analysis and Modeling of Complex Data in Behavioral and Social Sciences., pp. 201–209. Springer, Heidelberg (2014)

Evaluating the Quality of Data Imputation in Cardiovascular Risk Studies Through the Dissimilarity Profile Analysis

Nadia Solaro

Abstract Missing data handling is one of the crucial problems in statistical analyses, and almost always is overcome by imputation. Although the literature is rich in different imputation approaches, the problem of the assessment of the quality of imputation, i.e., appraising whether the imputed values or categories are plausible for variables and units, seems to have received less attention. This issue is critical in every field of application, such as the medical context considered here, i.e., the assessment of cardiovascular disease risks. We faced the problem of comparing the results obtained with different imputation methods and assessing the quality of imputation through the dissimilarity profile analysis (DPA), which is a multivariate exploratory method for the analysis of dissimilarity matrices. We also combined DPA with the traditional profile analysis for data matrices in order to improve understanding of the differentiation components among imputation methods.

Keywords Euclidean distance · Level · Missing data · Scatter · Shape

1 Introduction

Missing data handling is one of the crucial problems in statistical analyses, especially in the presence of multidimensional data. Almost always, handling of missing data is accomplished through imputation. A datum not available for any reason is replaced, by a suitable imputation method, with a value, if variables are quantitative, or an attribute/modality, if variables are categorical. The statistical literature is rich in a multitude of different imputation approaches (e.g., non-parametric vs parametric imputation methods, single vs multiple imputation methods) [5, 7]. Fewer efforts seem, however, to have been addressed to the problem of how to assess the quality of imputation (QoI), i.e., how to establish whether imputed data are

N. Solaro (✉)
Department of Statistics and Quantitative Methods, University of Milano-Bicocca, Milano, Italy
e-mail: nadia.solaro@unimib.it

F. Greselin et al. (eds.), *Statistical Learning of Complex Data*,
Studies in Classification, Data Analysis, and Knowledge Organization,
https://doi.org/10.1007/978-3-030-21140-0_9

consistent with the main features of variables and/or objects (or statistical units). The concept of QoI is undoubtedly related to the field of application. Nonetheless, two intertwined issues can be shared by many contexts: (1) the imputation plausibility for variables, i.e., detecting whether unrealistic values or categories have been imputed to incomplete variables—e.g., a negative value imputed to a variable that can assume only positive values—(*QoI for variables*); (2) the imputation plausibility for objects, i.e., assessing whether a datum imputed for an object is consistent with its profile as given by the values, or categories, or both, it has on the other variables (*QoI for objects*). We argue that these issues are critical in every field of application, all the more so in the specific medical context we considered, i.e., the assessment of cardiovascular disease (CVD) risks using information from the Autonomic Nervous System (ANS). In particular, we dealt with an overall dataset comprising variables collected from 88 different clinical studies undertaken over the period 1999–2014 and pertaining to specific groups of subjects (i.e., athletes, healthy individuals, smokers, stressed, obese and hypertensive subjects) [9, 12]. A missing data problem typically arises in contexts like this because of the adopted research protocols, which lay down rules for each clinical study. Depending on the protocol, measurements of some variables might not be contemplated within a specific study, especially if such information is too expensive or difficult to measure. In this sense, missing data appearing in the overall dataset can be regarded as generated by a MAR mechanism [12].

After having performed imputation on the overall dataset with different approaches, we faced the problem of comparing the results thus obtained by taking into account the consistency of the clinical group profiles against their expected traits. We then favoured the perspective of QoI for objects, instead of variables, although these two aspects are in strict relation. Comparisons among clinical group profiles imputed with the different methods were carried out by means of a multivariate exploratory tool for the analysis of dissimilarity matrices, i.e., the dissimilarity profile analysis (DPA) [8, 10], which here was also combined with the traditional profile analysis (PA) for multivariate data matrices [3] in order to deepen understanding of the differentiation components among profiles.

2 The DPA Method in Short

DPA is an exploratory multivariate statistical method designed for the analysis of dissimilarity matrices [8, 10]. It aims at investigating profiles of differences within the same set of objects to assess whether two objects differ to the others similarly or not, and then detect the main components that explain such differences. Given a set of n objects, let $\mathbf{\Delta}$ be an $(n \times n)$ square and symmetric matrix containing the dissimilarities: $\delta_{ir} = \delta_{ri}$, $i \neq r = 1, \ldots, n$, for which the usual properties hold [3]. In particular, matrix $\mathbf{\Delta}$ has a zero-diagonal for the self-dissimilarities ($\delta_{ii} = 0$ for all i). The elemental data of DPA are the dissimilarity profiles (DPs), which are given by the n row-vectors $\boldsymbol{\delta}_i^t = [\delta_{i1}, \delta_{i2}, \ldots, 0, \ldots, \delta_{i\,n-1}, \delta_{in}]$ of matrix $\mathbf{\Delta}$. Since

δ_i^t contains the estimates, according to any dissimilarity measure, of the degree of diversity of object i with respect to each of the other $n-1$ objects, the i-th DP is expression of the pattern of difference of object i, and can then be analysed to detect the underlying features that distinguish this DP in comparison with the other DPs. This kind of analysis relies on the decomposition of the squared Euclidean (EU) distance d_{ir}^2 between DPs [8, 10]: $d_{ir}^2 = \sum_{l=1}^{n}(\delta_{il} - \delta_{rl})^2 = d_{(ir)}^2 + 2\delta_{ir}^2$, $\forall i \neq r = 1, \ldots, n$, where $d_{(ir)}^2 = \sum_{\substack{l=1 \\ l\neq i\neq r}}^{n} (\delta_{il} - \delta_{rl})^2$ is the *leave-ir-out squared EU distance*, i.e., the squared EU distance between the i-th and r-th DPs computed excluding their direct comparison estimated by the term $2\delta_{ir}^2$. Similarly to PA [3], distance $d_{(ir)}^2$ can be decomposed in the three additive terms called DP level, DP scatter and DP shape components, respectively, according to the DP decomposition formula:

$$d_{(ir)}^2 = (n-2)(\tilde{\delta}_{i(r).} - \tilde{\delta}_{r(i).})^2 + (\tilde{\nu}_{i(r).} - \tilde{\nu}_{r(i).})^2 + 2\tilde{\nu}_{i(r).}\tilde{\nu}_{r(i).}(1 - \tilde{\theta}_{(ir)}), \tag{1}$$

for each $i \neq r$ [10]. The quantities appearing in (1) are:

- in the DP level component (first term in (1)), the *i-th DP leave-r-out level*: $\tilde{\delta}_{i(r).} = \frac{1}{n-2}\sum_{\substack{l=1 \\ l\neq i\neq r}}^{n} \delta_{il}$, which is the level of the i-th DP excluding itself and its reciprocal dissimilarity with the r-th DP. Clearly, it holds: $\tilde{\delta}_{i(r).} \geq 0$, for each $r = 1, \ldots, n$;
- in the DP scatter component (second term in (1)), the *i-th DP leave-r-out scatter*: $\tilde{\nu}_{i(r).}^2 = \sum_{\substack{l=1 \\ l\neq i\neq r}}^{n} (\delta_{il} - \tilde{\delta}_{i(r).})^2$, which gives the spread of the i-th DP around its leave-r-out level. It is a kind of deviance of the i-th DP around its level that is computed excluding itself and the r-th DP;
- in the DP shape component (third term in (1)), the *leave-ir-out shape of the pair (i, r) of DPs*: $\tilde{\theta}_{(ir)} = \frac{\tilde{\nu}_{(ir)}}{\tilde{\nu}_{i(r).}\tilde{\nu}_{r(i).}}$, where $-1 \leq \tilde{\theta}_{(ir)} \leq +1$, and $\tilde{\nu}_{(ir)} = \sum_{\substack{l=1 \\ l\neq i\neq r}}^{n} (\delta_{il} - \tilde{\delta}_{i(r).})(\delta_{rl} - \tilde{\delta}_{r(i).})$. Quantity $\tilde{\theta}_{(ir)}$ is a sort of correlation coefficient of the i-th and r-th DPs excluding their reciprocal dissimilarity. It measures the degree of similarity ($\tilde{\theta}_{(ir)}$ close to 1) or diversity ($\tilde{\theta}_{(ir)}$ close to -1) in the way that the two i-th and r-th DPs differ from the other DPs.

If also an $(n \times p)$ data matrix $\mathbf{X}$ is known, the usual PA can be applied and then combined with DPA. To apply standard results of PA, dissimilarities δ_{ir} in matrix $\boldsymbol{\Delta}$ have to be computed as EU distances of the n row-vectors $\mathbf{x}_i^t = [x_{ij}]_{j=1,\ldots,p}$ in $\mathbf{X}$ [3]. Then, the squared EU distance δ_{ir}^2 can be decomposed, in its turn, in the three additive terms representing each the P level, P scatter and P shape components:

$$\delta_{ir}^2 = p(\bar{x}_{i.} - \bar{x}_{r.})^2 + (v_i - v_r)^2 + 2v_i v_r(1 - q_{ir}), \qquad \forall i \neq r = 1, \ldots, n, \tag{2}$$

where $\bar{x}_{i.} = \frac{\sum_{j=1}^{p} x_{ij}}{p}$ is the level of the i-th profile; $v_i^2 = \sum_{j=1}^{p}(x_{ij} - \bar{x}_{i.})^2$ is the scatter of the i-th profile; $q_{ir} = \frac{v_{ir}}{v_i v_r}$ is the correlation coefficient of the i-th and r-th profiles, with $-1 \leq q_{ir} \leq +1$, and $v_{ir} = \sum_{j=1}^{p}(x_{ij} - \bar{x}_{i.})(x_{rj} - \bar{x}_{r.})$. To distinguish (2) from the DP decomposition (1), we denote formula (2) as the P decomposition. A critical requirement of PA is that variables have to be comparable regarding at least the unit of measurement. Otherwise, they have to be standardised [3].

3 DPA for Evaluating QoI in a CVD Risk Case Study

Analysis of QoI is considered within the case study related to the role of ANS proxies in the CVD risk assessment, as just mentioned in Sect. 1 and described in [12] and [9]. The overall dataset contains a total number of $n = 1314$ subjects and includes the variables shortly described in Table 1, which refer to personal data, ANS proxies, baroreflex gain index, blood pressure measures and body mass index. ANS proxies are the measures of the heart rate variability (or RR variability) resulting from the spectral analysis of the electrocardiogram traces, while the baroreflex gain index concerns the mechanism that helps blood pressure remain

Table 1 Definition of the variables considered in the case study

Sets of variables	Description
• Personal data	Age and gender (0 = Female, 1 = Male)
• Set A	*ANS proxies*:
($p_A = 7$ common	HR (heart rate, in beat/min)
complete variables)	RRMean (average of RR interval from tachogram, in msec)
	RRTP (total power, or RR variance, in ms^2)
	RRLFnu & RRHFnu (normalised power of low and high frequency
	components, resp., of RR variance, in nu)
	RRLFHF (ratio between absolute values of LF and HF)
	Anthropometrics:
	BMI (Body mass index: weight in kilos/height in m^2)
• Set B	*ANS proxies*:
($p_B = 8$ incomplete	RRLFHz & RRHFHz (centre frequency of LF and HF, resp., in Hz)
variables)	ΔRRLFnu (difference in LF power between stand and rest, in nu)
	Baroreflex gain index:
	AlphaM (frequency domain measure of baroreflex gain,
	in ms/mmHg)
	Blood pressure measures:
	SAP & DAP (systolic and diastolic arterial pressure, resp.,
	by sphygmomanometer, in mmHg)
	SAPMean (SAP average of systogram, in mmHg)
	SAPLFa (absolute power of LF component of systogram, in
	$mmHg^2$)

stable [6]. Apart from age and gender, wholly collected, the other $p = 15$ variables were divided into: (1) Set A, with the $p_A = 7$ common complete variables listed in Table 1, and (2) Set B, with the $p_B = 8$ incomplete variables reported in Table 1. On the other hand, $n_C = 836$ subjects have all the information (complete subjects), while the other $n_I = 478$ subjects have missing values in set B (incomplete subjects).

Imputation was carried out on the overall dataset through a multitude of different approaches [12]: (1) Non-parametric single imputation methods—IPCA (iterative principal component analysis) and FAMD (factorial analysis for mixed data) [4], FIP, FIM and WG.FIP (forward imputation) [11]; (2) Parametric multiple imputation methods—EM algorithm [2], MIPCA (multiple imputation with the PCA) [4]; (3) Data fusion non-parametric methods—hot-deck-distance-based methods, such as distance hot-deck (NND) and random hot-deck techniques (RndNND) [1]. These various imputation methods had given rise to different imputed ANS profiles of the clinical groups. The crucial point was therefore to choose the imputation method that best met the expected within-group ANS profiles according to the prior knowledge we had of the group features. To this end, we applied DPA and PA to compare the various imputed ANS profiles obtained for the same group and assess QoI according to the strategy of analysis described in Sect. 3.1.

3.1 Application of DPA and PA for Evaluating QoI

Before applying DPA and PA, we had to take into account that the groups were not directly comparable by age and gender. Inspections of the imputed ANS profiles were then necessarily based on variables adjusted for age and gender effects [9, 12]. The main traits of the groups were accordingly summed up by the adjusted median (AdjMed) profiles [12], which are of three types: (1) AdjMed profiles of the complete part ("benchmark profiles"), given by the within-group medians of the variables in sets A and B adjusted and standardised within the complete part of the data (i.e., the $n_C = 486$ subjects); (2) AdjMed profiles of the incomplete part, computed on the set of the incomplete subjects as within-group medians of the variables in set A adjusted and standardised within the entire set of the $n = 1314$ subjects; (3) AdjMed imputation (AdjMedImp) profiles, computed on the set of the incomplete subjects as within-group medians of the variables in set B adjusted and standardised within the whole set of the $n = 1314$ subjects.

We used these latter AdjMedImp profiles as input data of both PA and DPA, which we carried out within the groups separately considered. In particular, regarding PA, the input data matrix $\mathbf{X}_g$ of clinical group g contains, in its rows, the $n_g = 9$ AdjMedImp profiles, each referred to a specific imputation method, while, in its columns, the $p_B = 8$ variables of set B (Table 1). Values in $\mathbf{X}_g$ are thus given by the within-group-g medians of the eight adjusted and standardised variables imputed by the nine methods. Regarding DPA, its input data are given by the DPs $\delta^t_{i,g}$ of the

AdjMedImp profiles within group g, i.e., the rows of matrix $\mathbf{\Delta}_g$ whose elements are the EU distances δ_{ir} between the rows of matrix $\mathbf{X}_g$, $i \neq r = 1, \ldots, 9$.

For each clinical group, we computed the DP and P decompositions described in Sect. 2, formulas (1) and (2), and displayed the percentage results in the so-called complete DP decomposition plot. This plot is said as complete because it also includes the P decomposition (2). Besides, we referred to two further graphs to catch between-profiles differences better. The first graph, related to PA, is the profile plot, which depicts the above three types of AdjMed profiles against the variables in sets A and B. The second graph is the DP plot, which displays the normalised DPs of the imputation methods, i.e., the normalised EU distances of the AdjMedImp profiles: $\ddot{\delta}_{ir} = \frac{\delta_{ir}}{\max_{l \neq s}(\delta_{ls})}$, $\forall i \neq r$, since they help better detect the pairs of methods that are much more different/similar to the others.

We then assessed QoI within each clinical group by examining the results of PA and DPA jointly and bearing in mind our prior knowledge of the main traits of the incomplete subjects belonging to the various groups.

3.2 Assessment of QoI in the Athlete Group

Given the richness of the obtained results, we are going to refer to the athlete group only. The upper panel of Fig. 1 (see also [9]) contains the profile plot of the three types of AdjMed profiles against the variables in set A (first seven) and set B (last eight). Incomplete athletes (thin dashed line) have higher (median) values of BMI, RRMean and RRHFnu, and lower (median) values of HR, RRLFnu and RRLFHF than the complete athletes (benchmark profile, thin solid line). That is consistent with what expected because incomplete athletes were known to have more powerful traits than the complete ones. Accordingly, regarding the imputed variables (set B), higher values of ΔRRLFnu and AlphaM are expected along with potentially lower values of RRHFHz and SAPMean [12]. The AdjMedImp profiles of each method over set B (black or grey lines with different styles) have therefore to be compared taking into account the trend drawn for the incomplete part. While all the imputation methods appear to give consistent results for ΔRRLFnu, only the methods IPCA, FAMD, WG.FIP, MIPCA and EMlogtr produce the expected higher values for AlphaM. EMlogtr, however, produces too high levels of SAPMean, while both EMlogtr and MIPCA impute too low values for RRLFHz and RRHFHz. IPCA, FAMD and WG.FIP then appear to produce more plausible imputed ANS profiles.

Now, the crucial questions are: How similar/different are the results obtained in median by the considered imputation methods? Regarding which components do the AdjMedImp profiles mainly differ from one another? DPA was applied to address these questions and detect "different patterns of difference" between the methods. The lower panel of Fig. 1 reports the DP plot with the normalised DPs of the imputation methods (Sect. 3.1). Given the presence of the self-dissimilarities, each trajectory falls to zero in correspondence with the method to which it refers.

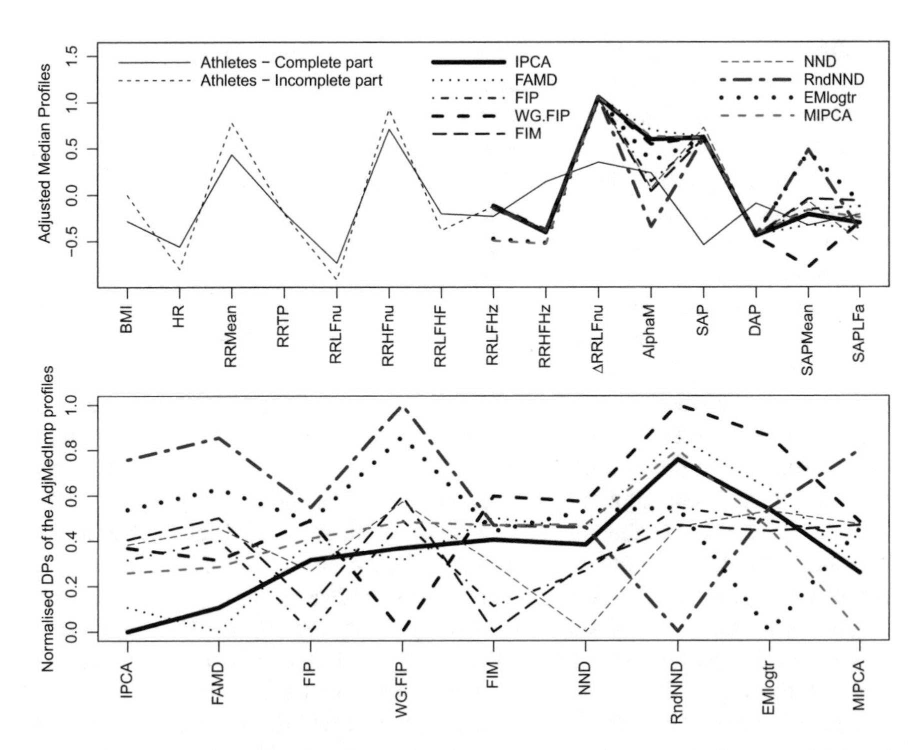

Fig. 1 Profile and dissimilarity profile analysis for the athlete group. Upper panel: Profile plot of adjusted median profiles (sets A and B). Lower panel: DP plot of the nine imputation methods

More importantly, the DP plot admits a double reading, in both a pairwise and a DP comparison perspectives. In a pairwise comparison perspective, by fixing a method on the horizontal axis, one can check how far the trajectories of the other techniques are to the zero, i.e., to the fixed method. The higher (lower) the trajectories are, the more different (similar) the other methods are in comparison with the one fixed on the horizontal axis. For instance, WG.FIP (bold dashed line) and RndNND (grey two-dashed line) produce the most different median imputation results ($\ddot{\delta}_{47} = \ddot{\delta}_{74} = 1$), while IPCA (bold solid line) and FAMD (thin dotted line) the most similar median imputation results ($\min_{i \neq r} \ddot{\delta}_{ir} = \ddot{\delta}_{12} = \ddot{\delta}_{21} = 0.106$). On the other hand, in a DP comparison perspective, DPs have to be analysed as whole trajectories. To compare a method DP with another one, we have to discard the part of the graph concerning their self-dissimilarities and reciprocal dissimilarity $\ddot{\delta}_{ir}$. In such a way, we can see how two different DP trajectories are over the other methods and thus ascertain whether the two methods share a similar pattern of difference compared to the others. For instance, the DP trajectories of WG.FIP and FAMD appear to be almost parallel, suggesting a substantial difference in the DP level component.

More thorough comparison analyses between trajectories are based on the DP and P decompositions (Sect. 2) of the squared EU distances d_{ir}^2 of the imputation method DPs. Figure 2 reports the complete DP decomposition plot concerning the percentage results of DPA (first row of panels) combined with PA (second row). Each of the components (DP level, DP scatter and DP shape, along with P level, P scatter and P shape) is depicted in a square plot. In the upper part of each plot, there are squares (DP decomposition) or circles (P decomposition) having areas proportional to the corresponding percentage recorded in the lower part. Over the six square plots, the percentages in the same position (i, r), with $i > r$, sum to 100%. In this way, for each pair of methods, we can see which components among the six best explain the observed patterns of difference. As an example, we have just noted that WG.FIP and RndNND have the most different median imputation results. By Fig. 2, their reciprocal dissimilarity represents the crucial component of difference, i.e., the P component (amounting in total to 74.84%), rather than their DP component (amounting in total to 25.16%). In other words, they mostly differ reciprocally in the way they impute values to variables in set B rather than in the way they differ from the other methods. For a better understanding, Fig. 3 provides two pairwise plots, the first for the P component, the second for the DP component. Regarding the reciprocal dissimilarity (P component), the upper panel contains the pairwise profile plot of WG.FIP and RndNND, which is obtained from the profile plot in Fig. 1 (upper panel) using only the variables in set B. The two trajectories differ quite exclusively in shape (74.34%, Fig. 2), i.e., WG.FIP and RndNND tend to impute similar values in median but with the exceptions of AlphaM and SAPMean. Regarding the DP component, the lower panel in Fig. 3 shows the pairwise DP plot of WG.FIP and RndNND, which is taken from the DP plot (Fig. 1, lower panel) by removing their self-dissimilarities and reciprocal dissimilarity. Also in this case, the two trajectories differ quite exclusively in shape (22.14%, Fig. 2) according to an opposite trend ($\tilde{\theta}_{47} = \tilde{\theta}_{74} = -0.67$). That means that WG.FIP and RndNND differ from the other methods by a different pattern. Where in the pairwise DP plot the WG.FIP trajectory is low (high), thus intending its similar (different) performance to some methods, the RndNND trajectory turns out to be high (low), thus denoting its different (similar) performance to those same methods.

Finally, returning to IPCA, FAMD and WG.FIP, which were just recognised as good candidates for imputation in the athlete group, by Fig. 2, they mostly tend to differ from one another for the DP component rather than the P component. In particular, the profile plot in Fig. 1 (upper panel) shows that they have similar profiles, which mainly differ in the P shape component for the way in which they impute values of AlphaM and SAPMean. Regarding the DP component, IPCA, FAMD and WG.FIP tend to differ similarly to the other methods, i.e., they share a similar pattern of difference because the DP level component weighs more than DP scatter and DP shape. In other words, imputation of IPCA, FAMD and WG.FIP differs similarly from the other methods for the magnitude of the median of the imputed values. The final choice among IPCA, FAMD and WG.FIP is also based on the third quartile of the adjusted imputation (Adj-Q3-Imp) profiles. The upper panel of Fig. 4 provides the profile plot of the Adj-Q3-Imp profiles of IPCA, FAMD and

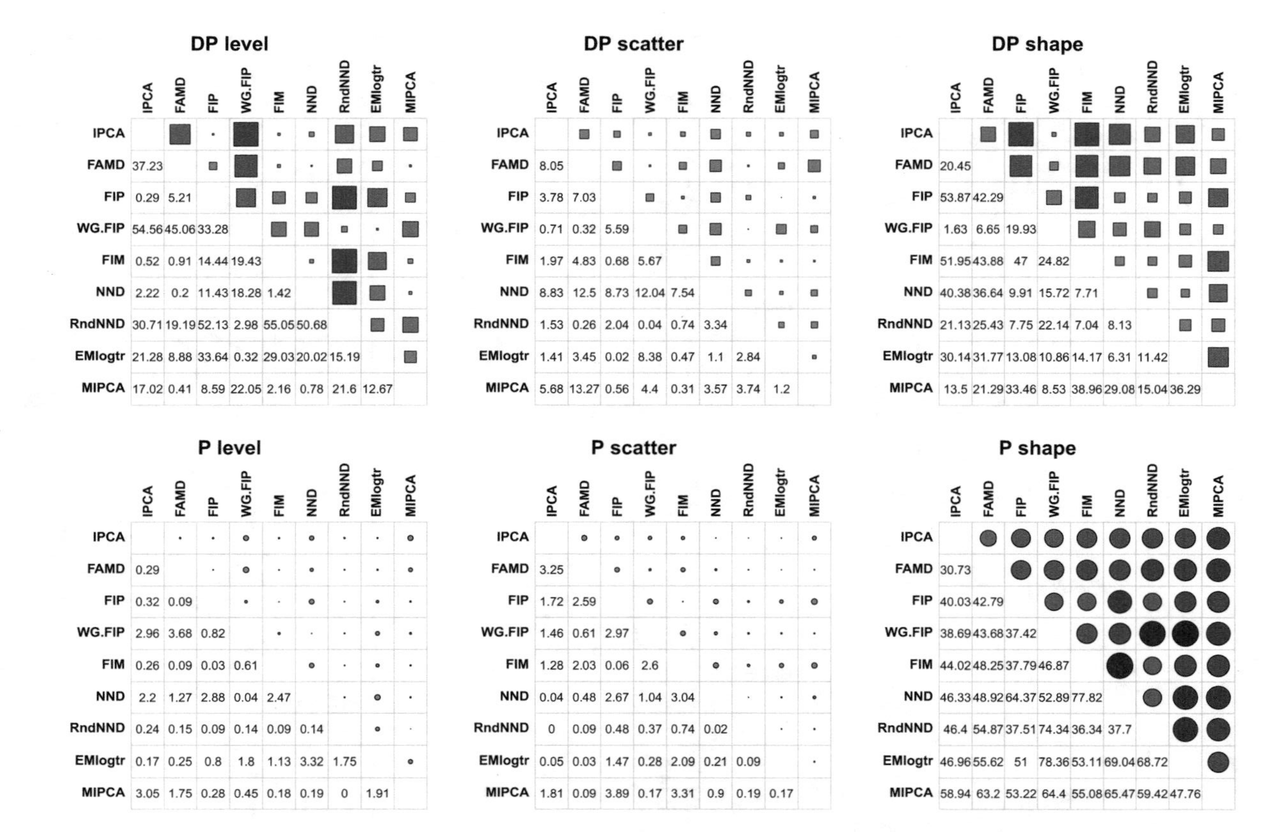

Fig. 2 Athlete group: Complete DP decomposition plot of the imputation methods. Upper three panels—DPA: DP decomposition into the DP level, DP scatter and DP shape components. Lower three panels—PA: P decomposition into the P level, P scatter and P shape components

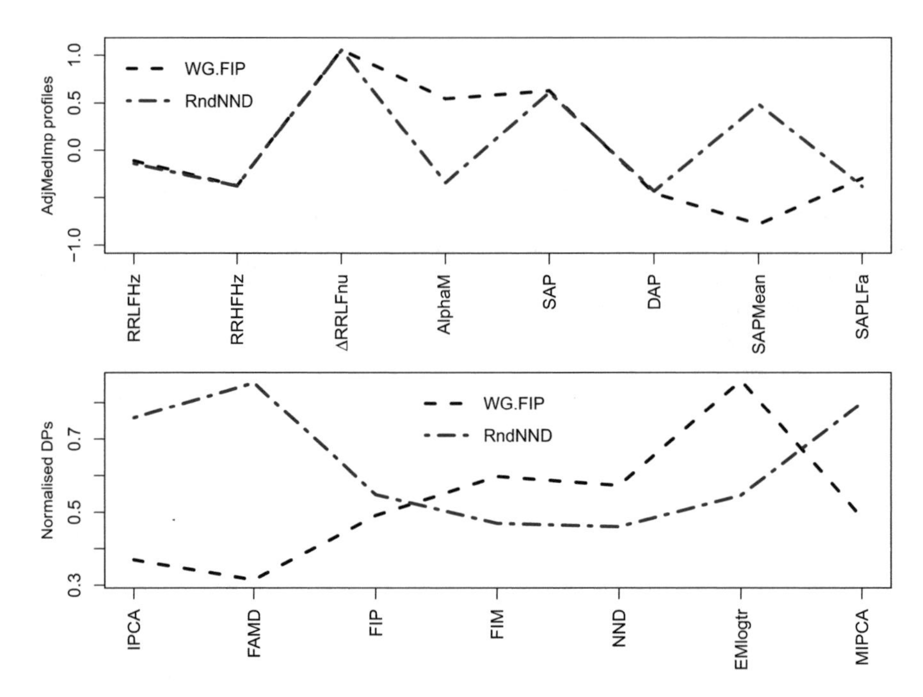

Fig. 3 Athlete group: Pairwise plots of WG.FIP and RndNND. Upper panel: Pairwise profile plot (P level: 0.14%, P scatter: 0.37%, P shape: 74.34%, Fig. 2). Lower panel: Pairwise DP plot (DP level: 2.98%, DP scatter: 0.04%, DP shape: 22.14%, Fig. 2)

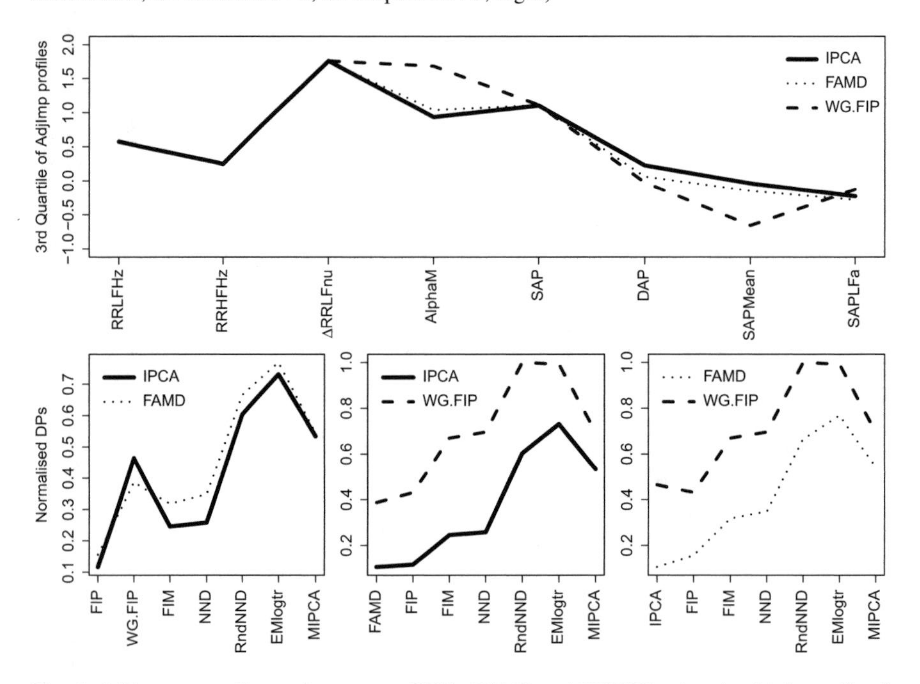

Fig. 4 Athlete group: Comparison among IPCA, FAMD and WG.FIP using the third quartile of the adjusted imputation profiles. Upper panel: Profile plot. Lower panels: Pairwise DP plots

WG.FIP, the lower panels the pairwise DP plots for the three pairwise comparisons. Similar remarks as those advanced for the AdjMedImp profiles can be made, with the unique exception that here IPCA and FAMD have the DP shape component higher than the DP level component (results omitted). As a final choice, WG.FIP is preferred to IPCA and FAMD because it guarantees higher imputed values of AlphaM and smaller imputed values of SAPMean also for the first 75% of the athletes.

4 Conclusions

Evaluation of quality of imputation (QoI) performed in the considered CVD risk assessment context was carried out by using two multivariate exploratory statistical methodologies, i.e., DPA and PA, in an integrated manner, and exploiting the available prior knowledge of the main features of the subjects having missing information within the clinical groups. From a practical point of view, one of the central facts emerged from the study was that the same imputation method might prove to be satisfactory for a clinical group, or a subset of subjects within it, but not for the other groups. For instance, the incomplete healthy subjects involved in specific clinical studies were known to be close to a hypertensive state. Compared with the set of the complete healthy subjects, lower levels of AlphaM along with higher values of blood pressure measures were then expected. The imputation method that reflected at best such a trend proved to be the distance hot-deck NND rather than WG.FIP as for the athletes. In general, a more suitable strategy to have realistic ANS imputed profiles could be that of switching from one imputation method to another depending on the features of the incomplete subjects, instead of applying a unique imputation method to the whole dataset. Such kinds of inspections would, however, require the availability of an integrated set of interactive diagnostic tools, capable of combining PA and DPA with the prior knowledge, when at hand, about missing features. The DPA method is at its early stage of development so that many aspects are still work-in-progress, e.g., the mathematical handling of DPs by interpolation or smoothing techniques. However, unlike the DP decomposition, the form of the DP trajectories is not invariant to the order in which the objects are taken. A primary challenge then is to set up procedures by which the objects can be arranged in a non-arbitrary order, or the horizontal axis of a DP plot can get metric properties.

Acknowledgements The author would like to thank Daniela Lucini and Massimo Pagani, BIOMETRA Department, University of Milan, for sharing their data and research on the neurovegetative system and CVD risk factors, and for their precious comments and suggestions.

References

1. D'Orazio, M., Di Zio, M., Scanu, M.: Statistical Matching - Theory and Practice. Wiley, New York (2006)
2. Honaker, J., King, G., Blackwell, M.: Amelia II: a program for missing data. J. Stat. Softw. **45**, 1–47 (2011)
3. Jobson, J.D.: Applied Multivariate Data Analysis. Volume II: Categorical and Multivariate Methods. Springer, New York (1992)
4. Josse, J., Pagès, J., Husson, F.: Multiple imputation in principal component analysis. Adv. Data Anal. Classif. **5**, 231–246 (2011)
5. Little, R.J.A., Rubin, D.B.: Statistical Analysis with Missing Data, 2nd edn. Wiley, New York (2002)
6. Lucini, D., Solaro, N., Pagani, M.: Autonomic differentiation map: a novel statistical tool for interpretation of heart rate variability. Front. Physiol. **9**, 401 (2018). https://doi.org/10.3389/fphys.2018.00401
7. Schafer, J.L.: Analysis of Incomplete Multivariate Data. Chapman and Hall/CRC, London (1997)
8. Solaro, N.: Dissimilarity profile analysis: a case study from Italian universities. Electron. J. Appl. Stat. Anal. **5**, 438–444 (2012)
9. Solaro, N.: Dissimilarity profile analysis for assessing the quality of imputation in cardiovascular risk studies. In: Greselin, F., Mola, F., Zenga, M. (eds.) Cladag 2017 Book of Short Papers, Universitas Studiorum S.r.l. Casa Editrice, Mantova, Italy (2017)
10. Solaro, N.: Dissimilarity profile analysis: a novel exploratory tool for dissimilarity matrices (2019, manuscript in preparation)
11. Solaro, N., Barbiero, A., Manzi, G., Ferrari, P.A.: A sequential distance-based approach for imputing missing data: forward imputation. Adv. Data Anal. Classif. **11**, 395–414 (2017)
12. Solaro, N., Lucini, D., Pagani, M.: Handling missing data in observational clinical studies concerning cardiovascular risk: an insight into critical aspects. In: Palumbo, F., Montanari, A., Vichi, M. (eds.) Data Science, Studies in Classification, Data Analysis, and Knowledge Organization Series, pp. 175–188. Springer International Publishing, Cham (2017)

Part III
Statistical Modeling

Measuring Economic Vulnerability: A Structural Equation Modeling Approach

Ambra Altimari, Simona Balzano, and Gennaro Zezza

Abstract Macroeconomic vulnerability is currently measured by the United Nations through a weighted average of eight variables related to exposure to shocks, and frequency of shocks, known as Economic Vulnerability Index (EVI). In this paper we propose to extend this measure by taking into account additional variables related to resilience, i.e., the ability of a country to recover after a shock. Since vulnerability can be considered as a latent variable, we explore the possibility of using the Structural Equation Model approach as an alternative to an index based on arbitrary weights. Using data from a panel of 98 countries over 19 years, we test our results with respect to the ability of the indices based on weighted averages, or on the SEM, in explaining the growth rate in real GDP per capita.

Keywords Hierarchical component model · Partial least squares · Structural equation modeling · Vulnerability index

1 Introduction

In the ongoing discussion on how to measure well-being and poverty, especially in relation to the allocation of international aid, the concept of *economic vulnerability* has emerged as potentially more useful than measures of poverty.

Several measures of vulnerability are used in the literature: they are mainly defined as *composite indicators*, typically computed as weighted averages of a set of indicators, where all indicators are assumed to have arbitrary (mostly equal) weights and to be uncorrelated to each other (i.e., correlation among them is ignored).

In order to identify countries that are eligible to enter or leave the Least Developed Countries category the United Nations refers, among other measures, to the Economic Vulnerability Index (EVI, [4]).

A. Altimari · S. Balzano (✉) · G. Zezza
Department of Economics and Law, University of Cassino and Southern Lazio, Cassino, Italy
e-mail: s.balzano@unicas.it; zezza@unicas.it

F. Greselin et al. (eds.), *Statistical Learning of Complex Data*,
Studies in Classification, Data Analysis, and Knowledge Organization,
https://doi.org/10.1007/978-3-030-21140-0_10

The EVI is computed as the average of two sub-indices: an *Exposure Index* and a *Shock Index*, which are weighted averages of five and three variables, respectively. As such, the EVI focuses on risk, but neglects measures of *resilience*, i.e., the ability of a country to recover after a shock.

We start our analysis from all EVI variables, observed on 98 developing countries between 1990 and 2013.[1]

Our contribution moves along two dimensions:

- we propose to extend the EVI model, including additional variables[2] affecting Resilience. We refer to this new specification as Extended–EVI, and compute its value as a weighted average of its determinants;
- we use the Structural Equation Model (SEM) approach to estimate the vulnerability index, both in its original specification and in our extended version. In this way, a weighting system deduced from the data replaces the fixed arbitrary weights for each year, and the correlation among (blocks of) variables plays its role in determining the vulnerability score. For this purpose we use the Partial Least Squares approach to Structural Equation Model (PLS–SEM) [5, 9, 11], whose aim is mainly predictive, to estimate the vulnerability index.

2 The Economic Vulnerability Index (EVI): A Possible Extension

The list of the base EVI variables is given in Table 1, the first five indicators measuring the exposure to risks and the last three referring to the outcomes of a shock.

All variables are expressed on a 0–100 scale and are measured so that a higher value implies higher vulnerability, i.e., so that they enter the index with positive weights.

The Extended–EVI includes nine additional indicators (listed in Table 2) affecting resilience. These new variables account for the *economic strength* (1–3) and the *strength of institutions* (4–9).

In the new specification we also include four additional variables for exposure to risk (see Table 2).[3] Additional variables have been selected for their theoretical relevance for vulnerability, subject to the availability of the data over time for a large enough number of countries.

[1]EVI data can be downloaded from http://byind.ferdi.fr/en/evi.

[2]Data sources: UN databases: UNSD–NA, UN–PD, UNCTAD Stat, FAOSTAT; World Bank databases: WDI, WGI; Centre for International Earth Science Information Network (CIESIN); Emergency Events Database (EM–DAT); Centre d'Etudes Prospectives et d'Informations Internationales (CEPII).

[3]For a detailed description of the variables, see the companion page at http://gennaro.zezza.it/files/abz.

Table 1 EVI variables [4]

1. Exposure	1.1 Population (smallness)
	1.2 Remoteness from world markets
	1.3 Export concentration
	1.4 Share of agriculture, forestry, and fisheries in GDP
	1.5 Share of population living in low elevated coastal zone
2. Shock	2.1 Victims of natural disasters
	2.2 Instability of agricultural production
	2.3 Exports instability

Table 2 The additional variables in the extended–EVI

3. Resilience	3.1 Net flows on external public and publicly guaranteed debt (%GDP)
	3.2 Debt service on external debt PPG (%GDP)
	3.3 Gross fixed capital formation (%GDP)
	3.4 Control of corruption
	3.5 Government effectiveness
	3.6 Political stability and absence of violence/terrorism
	3.7 Regulatory quality
	3.8 Rule of law
	3.9 Voice and accountability
1. Exposure	1.6 Surface area
	1.7 Import concentration
	1.8 Foreign direct investment net inflows (%GDP)
	1.9 Net official development assistance and official aid (%GDP)

3 The PLS Approach to Structural Equation Model

The PLS–Path Modeling (PLS–PM) [3, 9, 11], i.e., the component-based approach to SEM, is one of the most used estimation methods for latent variable models, such that the name PLS–SEM is often used as an alternative in the most recent literature [5]. It represents the main alternative to LISREL [6], differing, among other things, in the main aim being pursued, i.e., predictive (PLS–PM) versus confirmatory (LISREL), that makes it preferable for our aims.

Given a data matrix X, partitioned by column in J blocks, a *path diagram* (Fig. 1) is the typical representation of a causal model where each block X_j $(j = 1, \ldots, J)$ is a set of manifest variables and is conceptually connected to a latent variable ξ_j.

In such a diagram, rectangles represent manifest variables (MV), ellipses latent variables (LV), and arrows the relations between them, which are supposed to be linear. In Fig. 1 two models are shown, i.e., the base EVI model (in white) and the enlarged model including (in gray) the resilience and the additional variables for exposure.

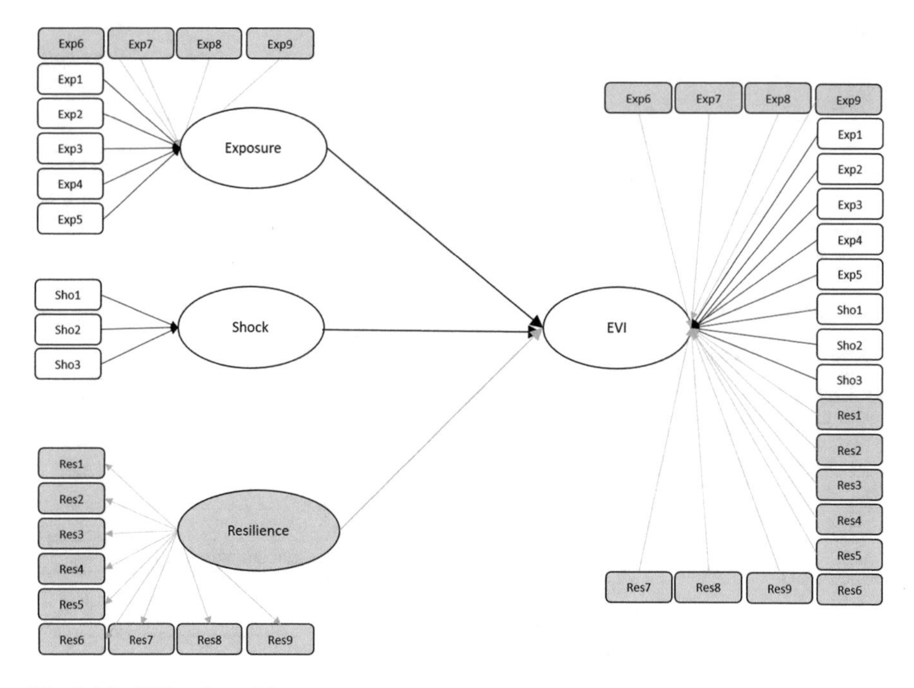

Fig. 1 The EVI path model

The PLS–PM algorithm computes latent variables scores, path coefficients (or inner weights), and outer weights. It is based on alternating, until convergence, an *external* and *internal* estimate of the LV, based on sets of regressions.

For details about the algorithms see [9] and [3].

We specify both the models for estimating the two versions of the EVI according to the Hierarchical Component Model, also referred to as Repeated Indicators Approach [7, 8, 10, 11], suitable to model structures with nested constructs. It is typically used for modeling composite indicators, when a set of sub-indices (first-order constructs) compound the global index (second-order construct) [5]. In such a model, all indicators in each of the J blocks are put together and used to define a $(J+1)$-th one, the so-called *super-block*, i.e., a higher-order latent variable whose final score is interpreted as the estimate of the composite indicator estimation. Thereby the vulnerability is estimated both as a linear combination of the manifest variables and as compound by sub-indices.

Moreover, we use consistent Partial Least Squares (PLSc) [1, 2] as estimation method, since the most recent literature suggests its adoption when some constructs are defined as factor models (as it is the case for the Resilience construct in our model).

The path model in Fig. 1 reproduces the conceptual structures of the two formulations of the index (EVI and Extended–EVI) in its basic specification. We specified this model according to the blocks' internal consistency. In particular we set Exposure and Shock, lowly internally correlated, as formative blocks and we add Resilience, which is unidimensional, as a reflective construct. On the structural part, the two (or three in the extended model) sub-indices are exogenous towards the Vulnerability, as conceptually Exposure and Shock (and Resilience in the extended model) can be seen as determinants of the global index.

4 Results

Results of this analysis consist in the estimation of 19 models, one for each year.

PLS–SEM assessment is based on measures indicating the model's predictive capability, mostly in terms of reliability and validity of the construct measures. Assessment of measurement models includes for each reflective block an *internal consistency reliability* index, measured by the Cronbach's alpha and assessing the unidimensionality (necessary condition to define a block as reflective), and a *convergent validity* index, measured by the average variance extracted (AVE).

For the block Resilience these indices range in the intervals [0.73; 0.79] and [0.41; 0.45] respectively, denoting a quite acceptable block structure.

A possible criterion to evaluate a formative block is the relevance of the indicators, to verify if they truly contribute to forming the construct. This is assessed by testing if the *outer weights*, i.e., the *relative contributions* of indicators, significantly differ from zero. And, since outer weights' significance can be affected by external factors, like the number of indicators in the block, it is convenient to consider also *outer loadings*, i.e., the *absolute contribution* of the indicator. In the estimated models bootstrap confidence intervals of outer weights are not significant in some cases, but most of the times they are associated to quite high and/or significant values of the outer loadings. Even if this condition is not always met (both across the 19 models and for all indicators), at an exploratory level we consider the model to be generally acceptable.

In order to evaluate separately the relevance of the extension of the EVI, and the adoption of a multivariate approach, we estimate two different models:

- the SEM–EVI: PLS–SEM estimation using the 8 base EVI indicators as manifest variables, related to 2 exogenous latent variables (Exposure and Shock) explaining an endogenous super-block (Vulnerability);
- the New–EVI: PLS–SEM estimation using the 21 indicators (8 base + the additional 13) as manifest variables, related to 3 exogenous latent variables (Exposure, Shock, and Resilience) explaining the endogenous super-block (Vulnerability).

We have also computed a fourth index, named Extended–EVI, as the average of our 21 indicators. Results obtained using different models/estimation methods are compared between each other over time at an empirical level. Table 3 shows the four indices, according to model specification and estimation method.

In the following we will start comparing the indices on the first row, to consider the consequences of just adding new variables to the base EVI, without changing

Table 3 The compared models

	Model specification	
Estimation method	8 variables	8+13 variables
(Weighted) Average	EVI	Extended–EVI
PLS–SEM	SEM–EVI	New–EVI

the estimation methods, and we will later compare the indices by the columns, to consider the consequences of using the SEM approach.

We will point out how the two indices perform in terms of (1) the coherence of the results with the underlying theoretical model and (2) the capability of the estimated indices to explain real GDP growth, which we choose as a simple aggregate measure of economic performance.

In synthesis, in showing our results we will refer to the following scheme:

1. To evaluate the internal coherence of the indices we mainly use some descriptive analysis of (1) signs and values of the estimated weights and their trend over time; (2) trend, correlations between and autocorrelations of indices; (3) countries' final rankings;
2. We next test the predictive power of the indices by regressing them on real GDP growth.

4.1 Comparing Different Models' Specifications

We compare EVI with Extended–EVI to observe how results change by simply adding the new 13 variables (9 for resilience plus 4 more for exposure) to the classical index. Correlation between the original measure and the enlarged measure is reasonably high (0.63).

Country by country comparison of the two indices show similar trends over time for most countries, albeit with exceptions. Comparing rankings obtained by the two indices, we notice that the countries with the largest change in their positions, such as Tonga or St. Kitts and Nevis, are those—as expected—characterized by the highest value for our resilience indicators.

4.2 Comparing Estimation Methods

As mentioned above, the main expected consequence of using SEM is to let the index weighting system emerge from the data. In most cases the outer weights of manifest variables on both SEM–EVI and New–EVI show basic instability. Moreover, the presence of some negative weights for some variables points out that the use of positive (and constant in time) weights in the classical EVI is not justified by the data. On the other hand, the assumption of positive weights does not rely on positive correlations between variables.

Table 4 Regression on GDP growth

	Growth in real GDP per capita			
Growth(t–1)	0.159[a]	0.148[a]	0.160[a]	0.156[a]
EVI	−0.020			
Extended–EVI		−0.162[a]		
SEM–EVI			0.053	
New–EVI				−0.047*
Intercept	2.73[a]	7.856[a]	0.144	3.809[a]
N	1706	1706	1706	1706
Adj R^2	0.37	0.38	0.37	0.37

Coefficients identified by (a) are significant at 1%. Coefficients identified by (*) are significant at 5%

The correlation between the official EVI and the computed SEM–EVI indices is generally low, ranging in $[-0.12; 0.22]$, and non-significant, while the two indices individually show a high autocorrelation from 1 year to the next, ranging in $[0.97; 0.99]$ (all significant values) for SEM–EVI and in $[0.72; 0.99]$ for New–EVI, (with the exception of year 2010 that is negatively correlated with all other years, values in $[-0.97; -0.77]$). For both autocorrelation is decreasing with distance in time, proving their strong internal coherence.

In other words, the two indices provide two different measures of vulnerability, each with its own ranking, but both have their own internal coherence. Extended–EVI is instead highly correlated to New–EVI, both for the whole sample and for many countries.[4]

4.3 Does Vulnerability Help Explain Growth?

In addition to the comparisons among indices, we have tested which of the proposed models performs better in explaining growth in real GDP per-capita, where vulnerability should have a negative impact on growth.[5] In Table 4 we report the results of four fixed-effects panel regressions on GDP growth for each of the four indices, adding lagged GDP growth as an additional explanatory variable.

We notice that the SEM–EVI has the "wrong" sign, while the two extended measures of vulnerability have the best performance, in terms of overall explanatory power. We control for five outliers: three associated to exceptional positive growth rates in GDP, and two related to large drops in GDP given by war episodes (Lybia 2011; Central African Republic, 2013)

[4]More detailed results are available at http://gennaro.zezza.it/files/abz.

[5]We are aware that this analysis cannot rule out the possibility that GDP growth has an impact on vulnerability, and that therefore our explanatory variables may not be weakly exogenous.

5 Final Remarks

We have analyzed the measure of economic vulnerability adopted by the United Nations, EVI, and proposed to extend it by considering additional indicators to take into account the ability of a country to recover from shocks. We have further proposed a multivariate approach, based on PLS–SEM, for estimating economic vulnerability indices. The proposed enlarged index performs better than the classical one in predicting the growth rate in real GDP per-capita, thus validating the usefulness and external coherence of our approach.

The multivariate approach has also shown that some of the manifest variables enter the EVI with low, in some cases negative, weights, casting doubts to the appropriateness of the EVI base model. Doubts are powered by the weakness of assessment measures provided by the basic and extended indices path models estimation.

As the analysis covers 98 countries on 19 years, we also investigated the possible role of time in repeated PLS estimates over the years. We consider that the relative stability of scores obtained in repeated PLS estimates to be reassuring in terms of model specification, but further results on applying the PLS method to a panel, fully exploiting the information contained in autocorrelation and cross-correlation of variables, will require further work.

References

1. Dijkstra, T.K., Henseler, J.: Consistent and asymptotically normal PLS estimator for linear structure equations. Comput. Stat. Data Anal. **81**, 10–23 (2015)
2. Dijkstra, T.K., Henseler, J.: Consistent partial least squares path modeling. MIS Q. **39**(2), 297–316 (2015)
3. Esposito Vinzi, V., Chin, W.W., Henseler, J., Wang, H.: Handbook of Partial Least Squares. Springer, Berlin (2010)
4. Guillaumont, P.: An economic vulnerability index: its design and use for international development policy. Oxf. Dev. Stud. **37**(3), 193–228 (2009)
5. Hair, J.F., Hult, T., Ringle, C.M., Sarstedt, M.: A Primer on Partial Least Squares Structural Equation Modeling (PLS-SEM). Sage, Thousand Oaks (2014)
6. Jöreskog, K.G., Sörbom, D., Magidson, J.: Advances in Factor Analysis and Structural Equation Models. Abstract Books, Cambridge (1979)
7. Tenenhaus, M., Esposito Vinzi, V.: PLS regression, PLS path modeling and generalized Procustean analysis: a combined approach for multi-block analysis. J. Chemometr. **19**(3), 145–153 (2005)
8. Tenenhaus, M., Hanafi, M.: A bridge between PLS path modeling and multi-block data analysis. In: Esposito Vinzi, V., et al. (eds.) Handbook of Partial Least Squares, pp. 99–109. Springer, Berlin (2010)
9. Tenenhaus, M., Esposito Vinzi, V., Chatelin, Y.M., Lauro, C.: PLS path modeling. Comput. Stat. Data Anal. **48**, 159–205 (2005)
10. Wetzels, M., Odekerken-SchrÃűder, G., van Oppen, C.: Using PLS path modeling for assessing hierarchical construct models: guidelines and empirical illustration. MIS Q. **33**(1), 177–195 (2009)
11. Wold, H.: Path models with latent variables: The NIPALS approach. In: Blalock, H.M., et al. (eds.) Quantitative Sociology, pp. 307–357 (1975)

Bayesian Inference for a Mixture Model on the Simplex

Roberto Ascari, Sonia Migliorati, and Andrea Ongaro

Abstract The Flexible Dirichlet (Ongaro and Migliorati, J. Multivar. Anal. 114:412–426, 2013) is a distribution for compositional data (i.e., data whose support is the simplex), which can fit data better than the classical Dirichlet distribution, thanks to its mixture structure and to additional parameters that allow for a more flexible modeling of the covariance matrix. This contribution presents two Bayesian procedures—both based on Gibbs sampling—in order to estimate its parameters. A simulation study has been conducted in order to evaluate the performances of the proposed estimation algorithms in several parameter configurations. Data are generated from a Flexible Dirichlet with $D = 3$ components and with representative parameter configurations.

Keywords Compositional data · Dirichlet mixture · Bayesian estimation · MCMC

1 Introduction

Some kind of data are defined on unusual mathematical spaces instead of classical ones as $\mathbb{R}^D$. For instance, compositional data belong to the D-dimensional simplex, defined as:

$$\mathscr{S}^D = \left\{ \mathbf{x} \in \mathbb{R}^D : x_i > 0, \sum_{i=1}^{D} x_i = 1 \right\}.$$

R. Ascari (✉) · S. Migliorati · A. Ongaro
Department of Economics, Management and Statistics, University of Milano-Bicocca, Milano, Italy
e-mail: roberto.ascari@unimib.it; sonia.migliorati@unimib.it; andrea.ongaro@unimib.it

F. Greselin et al. (eds.), *Statistical Learning of Complex Data*,
Studies in Classification, Data Analysis, and Knowledge Organization,
https://doi.org/10.1007/978-3-030-21140-0_11

This means that data $\mathbf{x}$ are positive vectors subject to a unit-sum constraint (i.e., proportions). Note that compositional data are prevalent in many disciplines (e.g., geology, medicine, economics, psychology, environmetrics, etc.); therefore, their proper treatment is a relevant issue. The Dirichlet is the most known distribution defined on the simplex. Although it has several mathematical properties, in many real applications it does not fit the data well, due to its extreme forms of simplicial independence or stiffness in cluster modeling.

In order to overcome these drawbacks, a new model for this type of data has been proposed [4]: the Flexible Dirichlet (FD). Since this model can be represented as a finite mixture with particular Dirichlet components, it follows that it is more adequate to capture cluster structure in data. In the literature an estimation procedure based on the EM algorithm already exists [3]. The aim of this contribution is to introduce a new parametrization for the FD distribution and to use it for developing a new Bayesian estimation procedure.

More precisely, in Sect. 2 we present the FD model and show some of its properties (i.e., the finite mixture structure and first and second moments). Then, in Sect. 3, we propose a first Bayesian procedure in order to estimate the parameters and point out its drawbacks. In order to overcome them, in Sect. 4 we propose a new parametrization that provides a variation independent parameter space, thus allowing to build up an efficient Gibbs sampling algorithm. Finally, in Sect. 5 we present a simulation study and show the results in two representative parameter configurations.

2 The Flexible Dirichlet Distribution

The FD distribution is obtained by normalizing a vector $\mathbf{Y} = (Y_1, \ldots, Y_D)$ with positive dependent elements, where $Y_i = W_i + Z_i U$, $(i = 1, \ldots, D)$, $W_i \sim \text{Gamma}(\alpha_i, \beta)$ are independent random variables (r.v.), $U \sim \text{Gamma}(\tau, \beta)$ is a further independent r.v., and $\mathbf{Z} = (Z_1, \ldots, Z_D) \sim \text{Multinomial}(1, \mathbf{p})$ is a random vector independent of W_i's and U. Let $Y^+ = \sum_{i=1}^D Y_i$; then the normalized vector $\mathbf{X} = \mathbf{Y}/Y^+$ is distributed as a FD$(\boldsymbol{\alpha}, \mathbf{p}, \tau)$ and its density function is:

$$f_{FD}(\mathbf{x}; \boldsymbol{\alpha}, \tau, \mathbf{p}) = \frac{\Gamma(\alpha^+ + \tau)}{\prod_{r=1}^D \Gamma(\alpha_r)} \left(\prod_{r=1}^D x_r^{\alpha_r - 1} \right) \sum_{i=1}^D p_i \frac{\Gamma(\alpha_i)}{\Gamma(\alpha_i + \tau)} x_i^\tau,$$

where $\mathbf{x} \in \mathscr{S}^D$, $\alpha^+ = \sum_{i=1}^D \alpha_i$, $\boldsymbol{\alpha} = (\alpha_1, \ldots, \alpha_D)$, $\alpha_i > 0$, $\tau > 0$, $0 \leq p_i < 1$, and $\sum_{i=1}^D p_i = 1$. Its distribution function can be written as a finite mixture with particular Dirichlet components:

$$FD(\mathbf{x}; \boldsymbol{\alpha}, \tau, \mathbf{p}) = \sum_{i=1}^D p_i \mathscr{D}(\mathbf{x}; \boldsymbol{\alpha}_i),$$

where $\mathscr{D}(\cdot\,;\,\cdot)$ denotes the distribution function of a Dirichlet r.v., $\boldsymbol{\alpha}_i = \boldsymbol{\alpha} + \tau \mathbf{e}_i$ and $\mathbf{e}_i$ is the vector whose elements are equal to 0 except for the ith element which is equal to 1. We recall the first two moments of this distribution:

$$\boldsymbol{\mu} = \mathrm{E}[\mathbf{X}|\boldsymbol{\alpha}, \tau, \mathbf{p}] = \frac{\boldsymbol{\alpha} + \tau\,\mathbf{p}}{\alpha^+ + \tau} = \frac{\boldsymbol{\alpha}}{\alpha^+}\left(\frac{\alpha^+}{\alpha^+ + \tau}\right) + \mathbf{p}\left(\frac{\tau}{\alpha^+ + \tau}\right)$$

$$\mathrm{Var}(X_i|\boldsymbol{\alpha}, \tau, \mathbf{p}) = \frac{\mathrm{E}[X_i|\boldsymbol{\alpha}, \tau, \mathbf{p}] \cdot (1 - \mathrm{E}[X_i|\boldsymbol{\alpha}, \tau, \mathbf{p}])}{\alpha^+ + \tau + 1} + \frac{\tau^2 p_i(1 - p_i)}{(\alpha^+ + \tau)(\alpha^+ + \tau + 1)}$$

$$\mathrm{Cov}(X_i, X_r|\boldsymbol{\alpha}, \tau, \mathbf{p}) = \frac{\mathrm{E}[X_i|\boldsymbol{\alpha}, \tau, \mathbf{p}] \cdot \mathrm{E}[X_r|\boldsymbol{\alpha}, \tau, \mathbf{p}]}{\alpha^+ + \tau + 1} - \frac{\tau^2 p_i p_r}{(\alpha^+ + \tau)(\alpha^+ + \tau + 1)}$$

$i, r = 1, \ldots, D,\ i \neq r$. Thanks to the mixture structure highlighted above, the density function of the FD can take on several shapes, including a number $k \leq D$ of different modes. Moreover, the FD can represent a good model for clustering, since it is a "structured" mixture with links among the component parameters, as it emerges from the definition of $\boldsymbol{\alpha}_i$. Each mixture component defines a cluster, whose vector mean is:

$$\boldsymbol{\theta}_i = \frac{\boldsymbol{\alpha} + \tau \mathbf{e}_i}{\alpha^+ + \tau}.$$

These cluster means deserve a very clear and simple geometric interpretation, as they are linear convex combinations of a common "barycenter" $\boldsymbol{\alpha}/\alpha^+$ and the ith simplex vertex $\mathbf{e}_i$. Thus, the ith element of $\boldsymbol{\theta}_i$ is higher than the ith element of $\boldsymbol{\theta}_j$, for every $j \neq i$. This introduces a very simple and reasonable form of differentiation among components, which is able to capture a broad range of cluster dissimilarities. The parameter $\frac{\tau}{\alpha^+ + \tau}$ measures the distance between each cluster mean $\boldsymbol{\theta}_i$ and the common barycenter $\boldsymbol{\mu}$ in direction of $\mathbf{e}_i$. Details about the E-M based procedure for obtaining the MLEs of the FD's parameters can be found in [3].

3 Bayesian Inference via Gibbs Sampling

First of all note that strong identifiability of the FD [4] ensures that this distribution does not show invariance under permutation of the mixture components. Therefore, no label switching problems arise in the estimation process.

In order to implement the Bayesian estimation procedure we need to define the likelihood function and the priors. Let $\mathbf{x} = (\mathbf{x}_1, \ldots, \mathbf{x}_j, \ldots, \mathbf{x}_n)$ be a sample of size n from $\mathbf{X} \sim \mathrm{FD}(\boldsymbol{\alpha}, \mathbf{p}, \tau)$, then the complete-data likelihood function can be written as:

$$L_C(\mathbf{x}, \mathbf{S}; \boldsymbol{\alpha}, \tau, \mathbf{p}) = \prod_{j=1}^{n}\prod_{i=1}^{D}\left\{p_i \frac{\Gamma(\alpha^+ + \tau)\Gamma(\alpha_i)}{\Gamma(\alpha_i + \tau)} x_{ji}^{\tau} \prod_{h=1}^{D} \frac{x_{jh}^{\alpha_h - 1}}{\Gamma(\alpha_h)}\right\}^{z_{ji}} \quad (1)$$

where z_{ji} is equal to 1 if the jth observation has arisen from the ith cluster of the mixture (i.e., $S_j = i$) and 0 otherwise [6].

As for prior elicitation, we can assume that $\mathbf{p}$ and $(\boldsymbol{\alpha}, \tau)$ have independent prior distributions. In this way we can choose a Dirichlet $(e_0, \ldots, e_0)$ prior for $\mathbf{p}$, where $e_0 \in \mathbb{R}^+$. This choice is coherent with the literature, where the Dirichlet with equal hyperparameters is the standard prior for the weights of a finite mixture model [1]. Another simple choice is to impose independence among τ and each α_i (i.e., $\pi(\boldsymbol{\alpha}, \tau) = \pi(\tau)\prod_{i=1}^{D}\pi(\alpha_i)$) and select a reparametrized exponential prior distribution for each element of the r.v. $(\alpha_1, \ldots, \alpha_D, \tau)$, which greatly simplifies computation of the full conditionals. Thus, we have:

$$\pi(\boldsymbol{\alpha}, \tau) \propto b^{\tau} \prod_{i=1}^{D} a_i^{\alpha_i}, \tag{2}$$

where $(a_1, \ldots, a_D, b)$ are positive hyperparameters.

Then, the Gibbs sampling implementation can be devised as follows. Let $\mathbf{S}$ denote the vector of missing group labels (i.e., $S_j = i$ means that the jth observation has arisen from group i). Then, the algorithm is composed by the following steps:

1. Obtain an initial classification $\mathbf{S}^{(0)}$ of data into D groups. Repeat steps 2 and 3 for $m = 1, \ldots, B, \ldots, B + N$.
2. Given $\mathbf{S}^{(m-1)}$, sample parameters from their full conditionals:
   - Sample $\mathbf{p}^{(m)}$ from $\pi(\mathbf{p}|\mathbf{S}^{(m-1)}, \mathbf{x})$
   - Sample $(\boldsymbol{\alpha}^{(m)}, \tau^{(m)})$ from $\pi(\boldsymbol{\alpha}, \tau|\mathbf{S}^{(m-1)}, \mathbf{x})$
3. Given the new parameters $(\boldsymbol{\alpha}^{(m)}, \tau^{(m)}, \mathbf{p}^{(m)})$, sample a new partition $\mathbf{S}^{(m)}$ from $\pi(\mathbf{S}|\boldsymbol{\alpha}^{(m)}, \tau^{(m)}, \mathbf{p}^{(m)})$

If we choose a Dirichlet prior for $\mathbf{p}$, then $\pi(\mathbf{p}|\mathbf{S}^{(m-1)}, \mathbf{x})$ has a Dirichlet distribution with parameters $(e_1, \ldots, e_D)$, where $e_i = e_0 + N_i(\mathbf{S}^{(m-1)})$ and $N_i(\mathbf{S}^{(m-1)})$ is the number of data points assigned to group i in partition $\mathbf{S}^{(m-1)}$. In order to obtain a new data partition $\mathbf{S}^{(m)}$, in step 3, we can generate a vector from a Multinomial $(1, \mathbf{p}_j^*)$ and assign to $S_j^{(m)}$ the position in which the 1 occur, where $\mathbf{p}_j^* = (p_{j1}^*, \ldots, p_{jD}^*)$ and:

$$p_{ji}^* = Pr\left(S_j = i|\boldsymbol{\alpha}^{(m)}, \tau^{(m)}, \mathbf{p}^{(m)}\right) = \frac{p_i^{(m)} f_{\mathscr{D}}(\mathbf{x}_j; \boldsymbol{\alpha}_i^{(m)})}{\sum_{k=1}^{D} p_k^{(m)} f_{\mathscr{D}}(\mathbf{x}_j; \boldsymbol{\alpha}_k^{(m)})}, \tag{3}$$

$(i = 1, \ldots, D)$ where $f_{\mathscr{D}}(\mathbf{x}_j; \boldsymbol{\alpha}_k)$ is the density function of a Dirichlet r.v. and $\boldsymbol{\alpha}_k = \boldsymbol{\alpha} + \tau\mathbf{e}_k$.

The main issue in this Gibbs sampling is the generation of values from the full conditional $\pi(\boldsymbol{\alpha}, \tau|\mathbf{S}^{(m-1)}, \mathbf{x})$. One can show that the latter represents a distribution difficult to generate from whatever prior we choose for $(\boldsymbol{\alpha}, \tau)$. Given the prior (2) we can compute the full conditionals:

$$\begin{cases} \pi(\alpha_l|\boldsymbol{\alpha}_{(-l)}, \tau, \mathbf{S}, \mathbf{x}) \propto \left[\dfrac{\Gamma(\alpha^+ + \tau)}{\Gamma(\alpha_l)}\right]^n \left[\dfrac{\Gamma(\alpha_l)}{\Gamma(\alpha_l + \tau)}\right]^{N_l(\mathbf{S})} a_l^{\alpha_l} \prod_{i=1}^{D} \prod_{j:S_j=i} x_{jl}^{\alpha_l} \\ \pi(\tau|\boldsymbol{\alpha}, \mathbf{S}, \mathbf{x}) \propto \left[\Gamma(\alpha^+ + \tau)\right]^n \prod_{i=1}^{D} [\Gamma(\alpha_i + \tau)]^{-N_i(\mathbf{S})} b^\tau \prod_{i=1}^{D} \prod_{j:S_j=i} x_{ji}^{\tau} \end{cases}$$

where $\boldsymbol{\alpha}_{(-l)} = (\alpha_1, \ldots, \alpha_{(l-1)}, \alpha_{(l+1)}, \ldots, \alpha_D)$.

Unfortunately, these full conditionals do not characterize some known distribution, so an inverse transformation method (ITM) has been implemented in order to obtain exact values from these distributions. This method requires the numerical evaluation of $D + 1$ integrals in order to compute the normalization constants for the full conditionals and one more numerical integration to obtain the distribution function of each one of the full conditionals. Finally, we have to numerically find the percentile associated with a value generated from a uniform distribution on (0, 1). This involves a time-consuming algorithm (i.e., slow convergence of the Gibbs sampler) though, as it has emerged from simulations we have implemented in R [5].

4 A New Parametrization

In order to overcome this drawback we propose the following new parametrization for the FD model:

$$\begin{cases} \boldsymbol{\mu} = \frac{\boldsymbol{\alpha}}{\phi} + \tilde{w}\mathbf{p} \\ \tilde{w} = \frac{\tau}{\phi} \end{cases} \qquad \begin{cases} \phi = \alpha^+ + \tau \\ \mathbf{p} = \mathbf{p} \end{cases}$$

This parametrization allows for an interesting interpretation of parameters: $\mathbf{p}$ are the usual weights of a mixture model, $\boldsymbol{\mu}$ represents the overall mean vector, ϕ is a precision parameter, and $\tilde{w}$ measures the distance of each cluster mean from the common barycenter $\boldsymbol{\mu}$. One can show that $\tilde{w} < \min_j \min\left\{\frac{\mu_j}{p_j}, 1\right\}$, so we can define a normalized version of $\tilde{w}$, i.e., $w = \dfrac{\tilde{w}}{\min_j \min\left\{\frac{\mu_j}{p_j}, 1\right\}}$. In this way the parameter space is variation independent, so that we can choose independent priors:

$$\begin{cases} \boldsymbol{\mu} \sim \mathscr{D}(e_0, \ldots, e_0) \\ w \sim \text{Unif}(0, 1) \end{cases} \quad \begin{cases} \phi \sim \text{Gamma}(g_1, g_2) \\ \mathbf{p} \sim \mathscr{D}(d_0, \ldots, d_0) \end{cases} \tag{4}$$

with e_0, d_0, g_1, and g_2 as positive hyperparameters. This set of priors ensures noninformativity, or at least vagueness, in the estimation procedure. Indeed, the

Dirichlet distribution with equal hyperparameters treats all the components alike. Moreover, the Gamma distribution is a common choice for the prior of the precision parameter, and, by choosing "small" values for the hyperparameters g_1 and g_2, vague priors are obtained. If we set $g_1 = g_2$ then there is a large prior probability on observed values all close to zero or one (as in this case the α_i's in the original parametrization would be less than 1). If this is not considered in tune with one's prior opinion, one might choose prior distributions for ϕ with higher mean, though still keeping a large variance (i.e., $g_1 = kg_2$, with $k \in \{10, 60, 100\}$ and $g_2 \in \{0.01, 0.001, 0.0001\}$). We implemented a Gibbs sampling algorithm by means of the BUGS software [2] in order to sample from the joint posterior distribution. The likelihood function used in this model is the complete-data likelihood function given by (1) written in terms of the new parameters. This is coherent with the Gibbs sampling structure described at the beginning of Sect. 3.

5 Simulation Study

In order to evaluate the performance of this Gibbs sampling algorithm, we simulated samples from a Flexible Dirichlet with $D = 3$ for several configurations of parameters. Priors as in (4) have been chosen with $e_0 = d_0 = 1$ and $g_1 = g_2 = 0.0001$. For space constraints, we report only the results for two representative parameter configurations: one with well separated clusters and one with overlapped clusters. The latter is a challenging scenario for every cluster-based approach, due to the difficulty to identify groups of homogeneous observations. In Fig. 1 we can see a simulated dataset for each of these scenarios.

We have generated 200 samples of size 150 for each parameter configuration and, for each of them, we initialized an MCMC chain of length 25,000 ($B = 10{,}000$

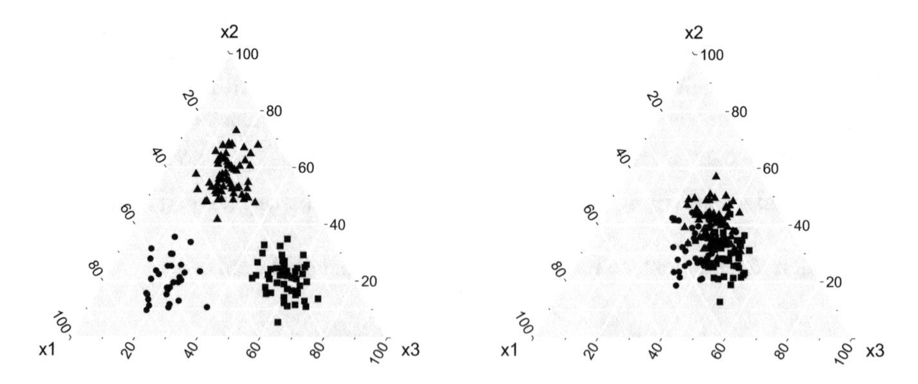

Fig. 1 Two dataset simulated from FD with: $\boldsymbol{\mu} = (0.333, 0.333, 0.333)$, $\mathbf{p} = (0.333, 0.333, 0.333)$, $\phi = 47$, $w = 0.362$ (left panel) and $\boldsymbol{\mu} = (0.271, 0.339, 0.390)$, $\mathbf{p} = c(0.333, 0.333, 0.333)$, $\phi = 58.5$, $w = 0.116$ (right panel)

Table 1 Simulation results for well separated clusters

Parameter	True	Post. Mean	Post. SD	MLE	MLE SE
μ_1	0.333	0.334	0.015	0.334	0.015
μ_2	0.333	0.335	0.015	0.335	0.015
μ_3	0.333	0.331	0.015	0.331	0.015
p_1	0.333	0.334	0.038	0.334	0.039
p_2	0.333	0.337	0.039	0.337	0.039
p_3	0.333	0.329	0.038	0.329	0.039
ϕ	47	47.237	3.872	47.824	3.827
w	0.3617	0.361	0.009	0.390	0.018

Table 2 Simulation results for overlapped clusters

Parameter	True	Post. Mean	Post. SD	MLE	MLE SE
μ_1	0.271	0.271	0.006	0.271	0.006
μ_2	0.339	0.340	0.006	0.340	0.006
μ_3	0.390	0.389	0.006	0.389	0.006
p_1	0.333	0.366	0.206	0.337	0.155
p_2	0.333	0.335	0.203	0.344	0.162
p_3	0.333	0.299	0.198	0.319	0.171
ϕ	58.5	48.684	8.720	59.332	9.950
w	0.1158	0.066	0.031	0.152	0.050

of which as burn-in) and, to properly treat autocorrelation, we set a thinning value equal to 10. Graphical tools (i.e., trace plots and mean plots) have been used in order to verify the convergence of the chain to the stationary distribution. In Tables 1 and 2 we have reported the mean of the 200 posterior means, the mean of the 200 posterior Standard Deviations (SD), the mean of the Maximum Likelihood Estimates (MLE) and of their Standard Errors (SE).

From Table 1 it emerges that, when clusters are well separated, our Bayesian procedure produces more accurate and less variable estimates than the E-M based ones. Nonetheless, if clusters are too closed together (see Table 2), both approaches do not provide an unbiased estimation of the parameters, as we expected due to the data structure. Though, in this scenario the classical procedure is preferable: the precision parameter ϕ and w are heavily underestimated with the Bayesian approach, while the ML procedure overestimates them only slightly.

Finally, note that our estimation procedure is robust with respect to the choice of the hyperparameters: even with different values of e_0, d_0, g_1, and g_2, we have obtained similar results as the ones showed in Tables 1 and 2. Furthermore, it is also robust with respect to the choice of the loss function: due to the approximate symmetry of each marginal posterior distribution, the posterior means are very close to the posterior medians and posterior modes.

In conclusion, we have introduced a new parametrization that allows to set up a more efficient Gibbs sampling algorithm. This new Bayesian method is very precise when data show separated clusters, but it does not work as well as the classical estimation procedure when clusters are overlapped.

References

1. Frühwirth-Schnatter, S.: Finite Mixture and Markov Switching Models. Springer, New York (1992)
2. Lunn, D.J., Thomas, A., Best, N., Spiegehalter, D.: WinBUGS-a Bayesian modelling framework: concepts, structure, and extensibility. Stat. Comput. **10**, 325–337 (2000)
3. Migliorati, S., Ongaro, A., Monti, G.S.: A structured Dirichlet mixture model for compositional data: inferential and applicative issues. Stat. Comput. **27**, 963–983 (2017)
4. Ongaro, A., Migliorati, S.: A generalization of the Dirichlet distribution. J. Multivar. Anal. **114**, 412–426 (2013)
5. R Core Team: R: A Language and Environment for Statistical Computing. R Foundation for Statistical Computing, Vienna. ISBN 3-900051-07-0. http://www.R-project.org/
6. Tanner, M.A., Wong, W.H.: The calculation of posterior distributions by data augmentation. J. Am. Stat. Assoc. **82**, 528–540 (1987)

Stochastic Models for the Size Distribution of Italian Firms: A Proposal

Anna Maria Fiori and Anna Motta

Abstract What determines the size distribution of business firms? What kind of firm dynamics may be underlying observed firm size distributions? Which candidate distributions may be used for fitting purposes? We here address these questions from a stochastic model perspective. We construct a firm dynamics process that leads to a Dagum distribution of firm size at equilibrium. An empirical study shows that the proposed model captures the empirical regularities of firm size distributions with considerable accuracy.

Keywords Firm dynamics · Gibrat's law · Dagum distribution

1 Introduction

The size distribution of business firms is one of the oldest and most relevant concepts in Industrial Organization studies. Knowledge of this distribution is fundamental for a number of reasons. Public policies often target small and mid-size firms whose growth is thought to have a beneficial effect in generating new employment opportunities. In contrast, the growth of large businesses is challenged by antitrust legislation due to questions of market power and unfair competition [2]. Managers operating in young (and hence often small) firms are aware that their companies must grow at a rapid pace to become productively efficient and survive. At different stages of their development, firms face different growth opportunities and must decide upon which of these opportunities should be taken up or discarded.

A. M. Fiori (✉)
Department of Statistics and Quantitative Methods, University of Milano-Bicocca, Milano, Italy
e-mail: anna.fiori@unimib.it

A. Motta
Il Sole 24 Ore, Milano, Italy
e-mail: anna.motta@unimib.it

F. Greselin et al. (eds.), *Statistical Learning of Complex Data*,
Studies in Classification, Data Analysis, and Knowledge Organization,
https://doi.org/10.1007/978-3-030-21140-0_12

The main message that has emerged from approximately first century of research in firm size distribution (FSD) is that the random element is prevailing across the growth experiences of firms [7]. For this reason, the literature has progressively shifted away from theoretical models that view firms as perfectly rational profit-maximizing entities. Attention is now focused on stochastic processes that capture the empirical regularities of the FSD, thus providing a more realistic description of the dynamics of the economic system (see, e.g., [1, 2, 7]).

In this work we contribute a new stochastic model of firm dynamics that leads to a Dagum distribution for the size of business firms operating in a given industry. The model builds on a stochastic growth process that was originally introduced in the context of income inequality studies [3, 6]. We here propose and empirically test an alternative parameterization. This sheds new light upon the connections between growth dynamics and the meaning of parameters that appear in the steady-state distribution of firm size.

The rest of the paper is organized as follows. In Sect. 2 we define a general stochastic framework for the study of firm growth and we introduce the Dagum distribution as a response to the main "stylized facts" about the FSD. In Sect. 3 we test the Dagum model on a dataset of Italian companies. Our findings and their implications are discussed in Sect. 4.

2 Model

Denote by $X(t)$ the size of an economic unit (firm) at time $t \geq 0$ and by $Y(t)$ its natural logarithm. A central mechanism for explaining the dynamics of $X(t)$ is multiplicative growth subject to random fluctuations [7]. This mechanism can be formulated by a doubly continuous (in time and states) stochastic process:

$$dY_t = g(y, t)dt + v(y, t)dB_t \tag{1}$$

where $g(y, t)$ is the infinitesimal drift coefficient, $v^2(y, t) > 0$ is the infinitesimal variance (diffusion coefficient), and $B(t)$ is a standard Brownian motion. Here, the drift term reflects the impact of deterministic forces on the instantaneous growth rate dY_t, while the Brownian motion fluctuations account for stochastic influences associated to uncertainty and risk factors [1].

The earliest and most influential model of type (1) was introduced in the 1930s by Gibrat [10], who viewed $Y(t)$ as the outcome of a large number of small additive influences, independent of each other and identically distributed with mean μ and variance σ^2. These assumptions imply that all firms in a given industry face the same distribution of growth rates independent of their size, a property that Gibrat called the *Law of Proportionate Effect*. Nesting Gibrat's Law into the general stochastic framework (1) gives an unrestricted Wiener process:

$$dY_t = \mu dt + \sigma dB_t. \tag{2}$$

This is a Gaussian process with $E[Y(t)] = \mu t$ and $Var[Y(t)] = \sigma^2 t$, both of which increase linearly with t (see, e.g., [12]). Hence Gibrat's Law implies a Lognormal distribution for the size of business firms, $X(t) = \exp[Y(t)]$, but this distribution is only transitional since its variance keeps growing over time.

Kalecki [11] observed that such growth was not characteristic of actual size distributions and in 1945 he amended Gibrat's model by postulating a negative correlation between growth rate and size. Based on the idea that large businesses face impediments to grow, Kalecki's model can be formulated as a mean reverting Ornstein-Uhlenbeck process [12] leading to a steady-state distribution for company size X that is Lognormal. Here, the Lognormal model emerges as a proper FSD and persists in the steady state as a consequence of "impeded growth."

The first applications of the Lognormal distribution were carried out by Gibrat and Kalecki themselves, and the goodness of fit they obtained for mid-size manufacturing establishments (respectively, in France and in the UK) was striking [14]. Starting in the 1980s, however, the availability of more complete datasets and the rise of computing technologies have gradually revealed the existence of a number of statistical regularities, or "stylized facts" that challenge the Gibrat-Kalecki model. In particular,

Fact 1 *Smaller firms grow relatively faster than their larger competitors.* In samples including small businesses, a negative relationship between firm size and expected growth has been repeatedly documented for manufacturing firms [2]. However, from a certain size onward, firms tend to experience constant returns to scale and thus have the same growth chances [14].

Fact 2 *Fluctuations in growth rates decrease with firm size.* Smaller firms tend to display a larger growth rate variance, which is plausible if one thinks that their basic structure is less diversified than that of big firms [7].

To incorporate these facts into the stochastic framework of Eq. (1), we introduce a generalized process of firm growth in which the drift and diffusion coefficients are explicitly modeled as functions of firm size. This process was originally discovered by Fattorini and Lemmi [6] in the context of income inequality studies and is reformulated here in a new parameterization.

Denote by $f(y) = lim_{t\to\infty} f(y,t)$ the steady-state density associated to the general stochastic process (1). If it exists, $f(y)$ satisfies the Forward Kolmogorov Equation:

$$0 = -\frac{\partial}{\partial y}[g(y)f(y)] + \frac{1}{2}\frac{\partial^2}{\partial y^2}\left[v^2(y)f(y)\right],$$

where $g(y) = lim_{t\to\infty} g(y,t)$ and $v^2(y) = lim_{t\to\infty} v^2(y,t)$. In accordance with Fact 1, we incorporate size dependence in the drift term by:

$$g(y) = -\frac{a\sigma^2}{2}\left[1 - (p-1)e^{-a(y-\log b)}\right] \tag{3}$$

where a, b, p, and σ are positive parameters. Based on Fact 2, the infinitesimal variance of Y is greatest at the lower end of the size scale. Hence, we specify the diffusion coefficient by:

$$v^2(y) = \sigma^2 \left[1 + e^{-a(y-\log b)}\right]. \tag{4}$$

Fattorini and Lemmi [6] have shown that this process has a steady-state density for Y which is a Type I Skew Logistic:

$$f(y) = ap \frac{e^{-a(y-\log b)}}{\left[1 + e^{-a(y-\log b)}\right]^{1+p}} \quad \text{for } y \in \Re, \tag{5}$$

with parameters $\log b$ (location), a (scale), and p (shape). It immediately follows that the equilibrium distribution of firm size, $X = \exp(Y)$, is given by a Dagum random variable with density:

$$f(x) = \frac{apx^{ap-1}}{b^{ap}\left[1 + \left(\frac{x}{b}\right)^a\right]^{p+1}} \quad \text{for } x > 0, \tag{6}$$

where a and p now play the role of shape parameters, and b is a scale (see, e.g., [3] and [13] for a detailed history of this multi-discovered distribution, its properties and its many parameterizations).

The Dagum density (6) is regularly varying at infinity with tail exponent $-a-1$ [13]. Thus smaller values of a imply that more probability mass is concentrated in the upper tail of the FSD. Interestingly, this can be related to the role of a as a scale parameter in the diffusion term (4): smaller values of a imply a higher instantaneous volatility of growth rates, particularly in the lower end of the size range (cf. Fact 2 above). This leads to a higher probability that firms grow to very large sizes.

The shape parameter p affects the skewness of the steady-state density of Y in connection with the behavior of the drift term (3). For $p = 1$, the drift is constantly equal to $-a\sigma^2/2$ and the Type I Skew Logistic (5) reduces to a conventional, symmetric Logistic density. This situation corresponds to a weaker form of Gibrat's Law in which the limiting mean of the (infinitesimal) growth rate is independent of firm size and negative (reflecting a *stability condition* explained, e.g., in [7]). Values of $p > 1$ imply a bounded, monotonic drift function that approaches $-a\sigma^2/2$ from above as y increases. The corresponding density for Y is positively skewed, reflecting a tendency of smaller firms to grow on average faster than their larger counterparts (cf. Fact 1 above).

3 Results of Empirical Studies

We tested the Dagum distribution on a 6-year panel of annual observations of total assets (size variable) of Italian companies operating in the Information and Communication Technologies (ICT) industry.[1] The industry includes NACE Rev. 2 Divisions 61 (telecommunications), 62 (computer programming, consultancy, and related activities), and 63 (information service activities), where NACE Rev. 2 is a classification of economic activities in the European Union managed by Eurostat.

Previous studies of the size distribution of ICT firms in Italy [8] rejected the hypothesis of lognormality due to a regularly varying upper tail. It is consequently worth investigating whether the Dagum distribution could be a suitable candidate for fitting purposes. Our analysis is illustrated for the logarithmic size variable Y whose characteristics are easier to visualize. The implications for the absolute size variable $X = \exp(Y)$ are immediately deduced.

The $18,476$ companies in our dataset have minimum total assets of 1000 Euros and were active in every year of the sample period, from 2010 to 2015. The choice of concentrating on relatively long-lived firms is consistent with our use of stochastic models that focus on the steady state for a closed population of firms. However, as argued in [4], such models have also practical relevance for understanding industries with entries and exits as long as these cancel out approximately.

The boxplots in Fig. 1 summarize the basic year-by-year information about central tendency, spread and possible outliers in Y.

The Normal and Type I Skew Logistic distributions were fitted to the log-size variable Y by Maximum Likelihood (ML) on a year-by-year basis (Table 1), yielding parameter estimates that appear fairly stable over time. Focusing on the shape parameter p, we carried out a formal test of the null hypothesis that $p = 1$ (symmetry) against the one-sided alternative that $p > 1$ (positive skewness). The test led to a strong rejection of H_0 in all years of the sample period. In view of the role played by p in the drift coefficient (3), this finding may be interpreted as evidence of an inverse relationship between expected growth and size, in consequence of which the distribution of Y deviates significantly from symmetry towards a positively skewed shape. This is confirmed by a visual comparison of the Normal and Type I Skew Logistic curves with the empirical histogram of log-size data (Fig. 2).

The similarity/discrepancy between the reference models and the empirical distribution of Y has been formally tested by two goodness-of-fit statistics: the Kolmogorov-Smirnov (KS) and the Anderson-Darling (AD) test (see, e.g., [8]). As shown in Table 1, the Normal distribution is rejected at all plausible significance levels, whereas the Type I Skew Logistic provides a very accurate description of Y in every year of the sample period. In particular, the Type I Skew Logistic fits the upper tail of the log-size distribution significantly better, as shown in Fig. 3.

[1]Data source: Aida, http://aida.bvdinfo.com/.

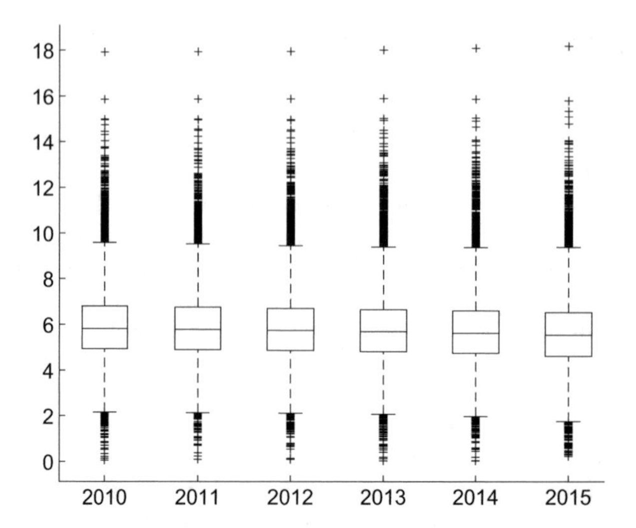

Fig. 1 Year-by-year boxplots of the logarithmic size variable Y for ICT firms. The empirical distribution of Y appears fairly stable over time, with a considerable number of points lying outside the whisker ends and a systematically longer whisker on top. These features suggest the presence of positive skewness and persistent fat tails that are unlikely to be compatible with a Normal distribution

A source of concern about these results is the impact of potential outliers represented by firms with total assets close to the minimum. On the one hand, small firms are the backbone of the Italian economy and nearly 10% companies in our dataset have total assets below 33,000 Euros. On the other hand, very small firms could have peculiar features (e.g. self-employment) that may lead to biased estimates. We carried out a sensitivity analysis by gradually removing fractions of very small firms from the dataset, respectively 0.5% (corresponding to firms with total assets below 6000 Euros) and 1% (total assets below 10,000 Euros). This led to a progressive increase in estimates of the skewness parameter p (Table 1, last two columns), suggesting that departures from normality in the log-size distribution are more pronounced when very small firms are excluded from the analysis.

4 Discussion

Ever since the work of Gibrat in 1930s, the Lognormal distribution has played a central role in studies of firm size distribution (FSD). A simple empirical test of this distribution consists in taking the natural logarithm Y of the firm size variable and comparing it to a Normal distribution. This comparison frequently reveals a poor fit, particularly in the tails (see, e.g., [9] for an interesting study of Gibrat's Law by Italian macro-regions).

Table 1 Normal and type I skew logistic distribution fit to the log-size of ICT firms: ML estimates of model parameters (standard errors in round brackets); goodness of fit tests (p-values in square brackets); sensitivity analysis excluding 0.5% (respectively 1%) of the smallest firms and computing a new estimate of the skewness parameter p' (resp. p'')

	Normal				Type I skew logistic					Sensitivity	
	ML estimates		Goodness of fit		ML estimates			Goodness of fit		0.5%	1%
Year	μ	σ	KS	AD	$\log b$	a	p	KS	AD	p'	p''
2010	5.628 (0.012)	1.576 (0.008)	0.0444[a] [0]	79.73[a] [0]	4.801 (0.053)	0.993 (0.011)	1.743[b] (0.061)	0.0051 [0.7159]	1.2850 [0.2373]	2.043[b] (0.082)	2.279[b] (0.100)
2011	5.743 (0.011)	1.532 (0.008)	0.0345[a] [0]	64.14[a] [0]	4.775 (0.053)	1.004 (0.010)	1.952[b] (0.071)	0.0054 [0.6532]	0.8756 [0.4296]	2.364[b] (0.101)	2.656[b] (0.125)
2012	5.807 (0.011)	1.521 (0.008)	0.0336[a] [0]	63.58[a] [0]	4.819 (0.053)	1.008 (0.010)	1.989[b] (0.072)	0.0052 [0.6890]	0.9056 [0.4108]	2.411[b] (0.104)	2.723[b] (0.130)
2013	5.858 (0.011)	1.514 (0.008)	0.0435[a] [0]	85.62[a] [0]	4.830 (0.054)	1.007 (0.010)	2.050[b] (0.076)	0.0051 [0.7139]	0.9193 [0.4025]	2.399[b] (0.103)	2.728[b] (0.130)
2014	5.903 (0.011)	1.516 (0.008)	0.0569[a] [0]	142.08[a] [0]	4.851 (0.056)	1.001 (0.010)	2.081[b] (0.079)	0.0052 [0.6950]	0.6441 [0.6073]	2.450[b] (0.107)	2.801[b] (0.138)
2015	5.934 (0.011)	1.529 (0.008)	0.0857[a] [0]	327.19[a] [0]	4.995 (0.054)	1.001 (0.010)	1.984[b] (0.073)	0.0048 [0.7899]	0.6379 [0.6129]	2.355[b] (0.100)	2.696[b] (0.129)

[a]The normal distribution is rejected at a significance level below 10^{-8}

[b]The skewness parameter p is significantly larger than 1 at (less than) 1% significance level, indicating that the best fit type I skew logistic density for Y is positively skewed

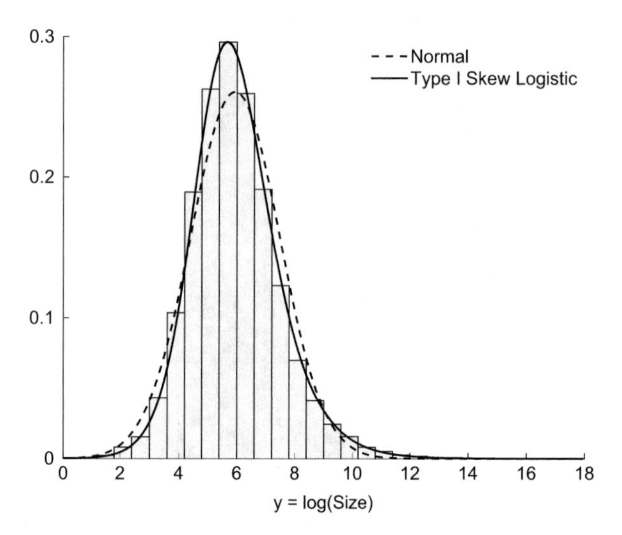

Fig. 2 Histogram of the log-size variable Y for year 2015: comparison with the normal and type I skew logistic curves fitted to data by maximum likelihood. The type I skew logistic fits the whole range of log-data with remarkable accuracy, capturing the presence of positive skewness and a heavy upper tail. The normal overestimates the frequency of small firms and underestimates the frequency of medium and large firms

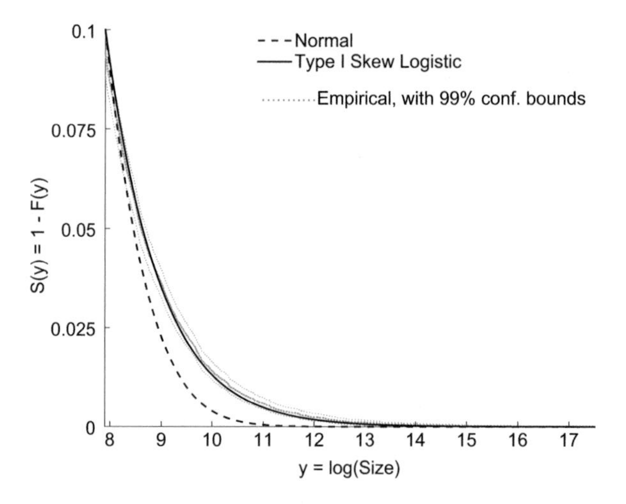

Fig. 3 Focus on the upper tail of the log-size variable Y for year 2015: comparison of the empirical survival function for the 10% largest firms with the normal and type I skew logistic curves. The type I skew logistic always lies inside the 99% confidence bounds for the empirical survival function, whereas the normal curve severely underestimates the probability that large businesses occur

As an alternative to the traditional assumption of normality for Y, we have proposed a stochastic model leading to a Type I Skew Logistic curve at equilibrium. This implies that company size, $X = \exp(Y)$, follows a Dagum distribution, a three parameter curve that combines an interior mode with a regularly varying upper tail. Our empirical study has shown that the Dagum model fits remarkably well the size distribution of Italian firms operating in the ICT industry.

As recently observed in [7], regular variation in the upper tail of the FSD may be interpreted as evidence that a weaker form of Gibrat's Law of Proportionate Effect holds for large businesses. In contrast with Kalecki's view, these businesses do not find impediments to grow in proportion to their size and, consequently, their growth opportunities are greater than a Lognormal distribution would predict. At the same time, the Dagum process implies that growth rate volatility is higher among smaller firms, which are nevertheless characterized by a tendency to grow on average faster than their larger competitors. In our study of Italian ICT firms, this tendency has been documented by results of statistical tests on the skewness parameter p that characterizes the drift of the Dagum process.

Although preliminary and limited to a specific dataset, our findings suggest potential implications for Industrial Organization studies and policy intervention. In particular, public policies targeting small and mid-size firms could be useful to (partly) offset their growth rate volatility, consolidate their growth patterns, and possibly help them maintain permanent employments [2]. This view seems to be shared also by the European Commission, which has recognized a prominent role of small and mid-size firms as drivers of economic growth [5]. On the other hand, the absence of impediments to grow (documented by evidence of a Paretian upper tail in the FSD) raises some questions as to whether very large businesses should be monitored to prevent possible industrial concentration processes [8]. These questions deserve further investigation. In particular, a more detailed study is in progress to test the Dagum model on different industries and to extend its basic formulation with a view on modeling industries with entries and exits of firms.

Acknowledgements We are indebted to Angiola Pollastri for her constant help and support, to Lisa Crosato for kindly sharing her experience with us, and to Simone Alfarano for his careful reading of an earlier draft that he contributed to improve with valuable comments. Helpful suggestions from two referees and from participants in CLADAG 2017 are gratefully acknowledged.

References

1. Alfarano, S., Milaković, M., Irle, A., Kauschke, J.: A statistical equilibrium model of competitive firms. J. Econ. Dyn. Control **36**, 136–149 (2012)
2. Coad, A.: The Growth of Firms: A Survey of Theories and Empirical Evidence. Edward Elgar, Cheltenham (2009)
3. Dagum, C.: A new model of personal income distribution: specification and estimation. Economie appliquée **30**, 413–437 (1977)
4. De Wit, G.: Firm size distributions: an overview of steady-state distributions resulting from firm dynamics models. Int. J. Ind. Organ. **23**, 423–450 (2005)
5. Eurostat: SME were the main drivers of economic growth between 2004 and 2006. Stat. Focus **71**, 1–8 (2009)

6. Fattorini, L., Lemmi, A.: Proposta di un modello alternativo per l'analisi della distribuzione personale del reddito. Atti delle Giornate di Lavoro AIRO **28**, 89–117 (1977)
7. Gabaix, X.: Power laws in economics and finance. Annu. Rev. Econ. **1**, 255–294 (2009)
8. Ganugi, P., Grossi, L., Crosato, L.: Firm size distributions and stochastic growth models: a comparison between ICT and mechanical Italian companies. Stat. Methods Appl. **12**, 391–414 (2003)
9. Ganugi, P., Grossi, L., Gozzi, G.: Testing Gibrat's law in Italian macro-regions: analysis on a panel of mechanical companies. Stat. Methods Appl. **14**, 101–126 (2005)
10. Gibrat, R.: Les inégalités économiques. Librairie du Recueil Sirey, Paris (1931)
11. Kalecki, M.: On the Gibrat distribution. Econometrica **13**, 161–170 (1945)
12. Karlin, S., Taylor, H.E.: A Second Course in Stochastic Processes. Academic Press, San Diego (1981)
13. Kleiber, C., Kotz, S.: Statistical Size Distributions in Economics and Actuarial Sciences. Wiley, Hoboken (2003)
14. Sutton, S.: Gibrat's legacy. J. Econ. Lit. **35**, 40–59 (1997)

Modeling Return to Education in Heterogeneous Populations: An Application to Italy

Angelo Mazza, Michele Battisti, Salvatore Ingrassia, and Antonio Punzo

Abstract The Mincer human capital earnings function is a regression model that relates individual's earnings to schooling and experience. It has been used to explain individual behavior with respect to educational choices and to indicate productivity on a large number of countries and across many different demographic groups. However, recent empirical studies have shown that often the population of interest embed latent homogeneous subpopulations, with different returns to education across subpopulations, rendering a single Mincer's regression inadequate. Moreover, whatever (concomitant) information is available about the nature of such a heterogeneity, it should be incorporated in an appropriate manner. We propose a mixture of Mincer's models with concomitant variables: it provides a flexible generalization of the Mincer model, a breakdown of the population into several homogeneous subpopulations, and an explanation of the unobserved heterogeneity. The proposal is motivated and illustrated via an application to data provided by the Bank of Italy's Survey of Household Income and Wealth in 2012.

Keywords Mincer's earnings function · Mixtures of regression models

1 Introduction

Earnings functions are used by social scientists to explain individual behavior with respect to educational choices and to indicate productivity [38]. They provide an indicator of returns to schooling, typically in the form of projected future wages,

A. Mazza · S. Ingrassia · A. Punzo (✉)
Dipartimento di Economia e Impresa, University of Catania, Catania, Italy
e-mail: a.mazza@unict.it; s.ingrassia@unict.it; antonio.punzo@unict.it

M. Battisti
Dipartimento di Scienze Giuridiche, della Società e dello Sport, University of Palermo, Palermo, Italy
e-mail: michele.battisti@unipa.it

F. Greselin et al. (eds.), *Statistical Learning of Complex Data*,
Studies in Classification, Data Analysis, and Knowledge Organization,
https://doi.org/10.1007/978-3-030-21140-0_13

which helps individuals to decide how to invest in their own human capital [39]. These indicators have also been used to relate fertility decisions with opportunity costs; since rearing of children is time intensive, an increase in earnings may induce a negative substitution effect on the demand for children [35, 49].

Introduced by Jacob Mincer, a pioneer of the New Labor Economics, in his seminal work Schooling, Experience and Earnings, the "human capital earnings function" is arguably the most popular earning function [34]. It is a single-equation model that explains the natural logarithm of earnings as a linear function of years of education, years of potential labor market experience, and the square of years of potential experience; in formula,

$$\ln(y) = \mu(\boldsymbol{x}; \boldsymbol{\beta}) + \varepsilon = \beta_0 + \beta_1 x_1 + \beta_2 x_2 + \beta_3 x_2^2 + \varepsilon, \tag{1}$$

where y denotes earnings, x_1 and x_2 represent years of education and years of potential labor market experience,[1] while $\varepsilon \sim N(0, \sigma)$, with σ being the (conditional) standard deviation of $\ln(y)$. In (1), $\boldsymbol{x} = (x_1, x_2)'$ and $\boldsymbol{\beta} = (\beta_0, \beta_1, \beta_2, \beta_3)'$.

The Mincer equation owes its popularity to the straightforward interpretation of the coefficient β_1 as approximated rate of return to education [8]. It has been examined on many datasets, involving a large number of countries and many different demographic groups and, as stated by Lemieux [28], it is "one of the most widely used models in empirical economics." Within income inequality studies, it has been used to study wage differentials due to gender [50] and for predicting the wage that a self-employed worker in a certain sector of the economy would have received on average as a paid employee on the same sector of economy [2, 3]. The literature on educational mismatches uses different specifications of (1) for quantifying the effect of educational mismatch on wages [37].

However, recent empirical studies [6, 23] have shown the relevance of unobserved heterogeneity. That is, often the populations of interest are constituted by latent groups bearing different returns to educations and characterized by different socio-demographic profiles, in a way that regression coefficients (and dispersion parameters) cannot be assumed to be the same for all observations, making the use of a single Mincer's regression inadequate. Heterogeneity and segmentation has been shown for Italy by Battisti and Cipollone [6, 14]. In fact, the Italian labor market has been traditionally characterized by rigid institutions, with employment protection legislation imposing strict rules and constraints regarding the ability to hire and fire workers. Reforms that started at the end of the 1990s have increasingly introduced more flexibility, but these new rules mostly apply only to newly hired workers; this has led to the formation of a two-tier labor market (see, e.g., [9]).

Finite mixtures of linear regression models, introduced by Quandt and Ramsey [48] in the general form of "switching regression," constitute a reference framework of analysis when no information about group membership is available and the modeling aim is to find groups of observations with similar regression

[1] See [12, 13] for a discussion on the use of polynomial terms for experience.

coefficients. Furthermore, whatever (concomitant) socio-demographic information is available about the nature of such a heterogeneity, it should be incorporated in the model in an appropriate manner.

To deal with these issues, in Sect. 2 we introduce a finite mixture of Mincer's regressions with concomitant variables: the proposal simultaneously provides a flexible generalization of the Mincer regression, a breakdown of the population into several homogeneous subpopulations, and an explanation of the unobserved heterogeneity also based on the considered concomitant variables. The EM algorithm is used for parameter estimation and BIC is adopted to select the number of subpopulations. In Sect. 4, the model is applied to disposable household income, as obtained from the Bank of Italy's Survey of Household Income and Wealth (SHIW) in 2012. In addition to illustrate the use of the model, this application demonstrates, based on the BIC, how a single Mincer's regression is inadequate for these data.

2 The Model

Given a d-dimensional vector $\boldsymbol{W}$ of concomitant variables (individual characteristics), and based on [16], we propose to generalize Eq. (1) via a finite mixture of k Mincer's regressions with concomitant variables; being a mixture, the proposed model can be defined from the conditional density of $\ln(y)$, given $\boldsymbol{x}$ and $\boldsymbol{w}$, in the following way

$$p\left[\ln(y)\,|\boldsymbol{x},\boldsymbol{w};\boldsymbol{\vartheta}\right]=\sum_{j=1}^{k}\pi_j\left(\boldsymbol{w};\boldsymbol{\alpha}\right)\phi\left[\ln(y)\,|\boldsymbol{x};\mu\left(\boldsymbol{x};\boldsymbol{\beta}_j\right),\sigma_j\right], \tag{2}$$

where $\pi_j(\boldsymbol{w};\boldsymbol{\alpha})$ are positive weights (depending on the parameters $\boldsymbol{\alpha}$) summing to one for each $\boldsymbol{w}$, $\phi(\cdot;\mu,\sigma)$ denotes the density of a Gaussian random variable with mean μ and standard deviation σ, $\mu(\boldsymbol{x};\boldsymbol{\beta}_j)$ is defined as in (1), and $\boldsymbol{\vartheta}$ contains all of the parameters of the model. The multinomial logit model

$$\pi_j\left(\boldsymbol{w};\boldsymbol{\alpha}\right)=\exp(\alpha_{j0}+\boldsymbol{\alpha}'_{j1}\boldsymbol{w})\Big/\sum_{h=1}^{k}\exp(\alpha_{h0}+\boldsymbol{\alpha}'_{h1}\boldsymbol{w}) \tag{3}$$

is assumed for the mixture weights in (2), where $\boldsymbol{\alpha}_{j1}=(\alpha_{j1},\ldots,\alpha_{jd})'$, $\boldsymbol{\alpha}_j=(\alpha_{j0},\boldsymbol{\alpha}'_{j1})'\in I\!R^{d+1}$, and $\boldsymbol{\alpha}=(\boldsymbol{\alpha}'_1,\ldots,\boldsymbol{\alpha}'_k)'$, with $\boldsymbol{\alpha}_1\equiv\boldsymbol{0}$ for identifiability sake [22, 24].

Model (2) can be used as a powerful device for clustering by assuming that each mixture component represents a group underlying the overall population [33]. Advantageously, by means of the mixture weights in (3), the concomitant variables $\boldsymbol{w}$ can be used to explain the profiles of the different groups. Here, it is important to stress that, based on the mixture model (2), we do not specify the groups a priori, but we let the data identify homogeneous groups with respect to the relationship between $\ln(y)$ and $\boldsymbol{x}$. For alternative uses of covariates and concomitant variables in mixtures of regressions models, see [7, 15, 25, 26, 32, 41–44, 46, 51, 52, 56].

3 Maximum Likelihood Estimation: The EM Algorithm

To find maximum likelihood (ML) estimates for $\boldsymbol{\vartheta}$ in (2), we adopt the EM algorithm of [17], as implemented by the `stepFlexmix()` function of the **flexmix** package [22] for R. In detail, given a random sample $(y_1, \boldsymbol{x}_1', \boldsymbol{w}_1')', \ldots, (y_n, \boldsymbol{x}_n', \boldsymbol{w}_n')'$ of $(Y, \boldsymbol{X}, \boldsymbol{W})$ from model (2), and once k is assigned, the algorithm basically takes into account the complete-data log-likelihood

$$l_c(\boldsymbol{\vartheta}) = \sum_{i=1}^{n}\sum_{j=1}^{k} z_{ij} \ln\left[\pi_j(\boldsymbol{w}_i; \boldsymbol{\alpha})\right] + \sum_{i=1}^{n}\sum_{j=1}^{k} z_{ij} \ln\left[p\left(y_i|\boldsymbol{x}_i; \boldsymbol{\beta}_j, \sigma_j\right)\right], \tag{4}$$

where $z_{ij} = 1$ if $\left(y_i, \boldsymbol{x}_i', \boldsymbol{w}_i'\right)'$ comes from component j and $z_{ij} = 0$ otherwise. The EM algorithm iterates between two steps, one E-step and one M-step, until convergence; their schematization, with respect to model (2), is given below (see [55, pp. 120–124] for further details).

E-step: Given the current parameter estimates $\boldsymbol{\vartheta}^{(r)}$ of the rth iteration, each z_{ij} is replaced by the estimated posterior probability

$$z_{ij}^{(r)} = \pi_j\left(\boldsymbol{w}_i; \boldsymbol{\alpha}^{(r)}\right) \phi\left[\ln(y_i)\,|\boldsymbol{x}_i; \mu\left(\boldsymbol{x}_i; \boldsymbol{\beta}_j^{(r)}\right), \sigma_j^{(r)}\right] \Big/ p\left[\ln(y_i)\,|\boldsymbol{x}_i, \boldsymbol{w}_i; \boldsymbol{\vartheta}^{(r)}\right]. \tag{5}$$

M-step: the obtained values of $z_{ij}^{(r)}$, which are function of $\boldsymbol{\vartheta}^{(r)}$, are substituted to z_{ij} in (4) so leading to the expected complete-data log-likelihood which is maximized with respect to $\boldsymbol{\vartheta}$, subject to the constraints on these parameters.

The EM algorithm described above needs to be initialized. Among the possible initialization strategies [4, 27], in the real data analysis of Sect. 4 a random initialization is repeated ten times from different random positions and the solution maximizing the observed-data log-likelihood among these ten runs is selected (see also [5, 29, 45, 47]).

Once the model is fitted, we can estimate the posterior probabilities (MAP) of group membership, say $\widehat{z}_{ij}$, based on (5). Hence, each individual can be assigned to one of the k groups via the *maximum a posteriori* probability operator $\text{MAP}\left(\widehat{z}_{ij}\right)$, assuming value 1 if $\max_{h=1,\ldots,k}\{\widehat{z}_{ih}\}$ occurs at component j and 0 otherwise.

Finally, to select the number of mixture components k, we adopt the Bayesian information criterion $\text{BIC} = -2l(\hat{\boldsymbol{\vartheta}}) + m \ln n$, where m is the overall number of parameters in the model and $l(\hat{\vartheta})$ denotes the maximized observed-data log-likelihood.

4 Real Data Analysis

We use data provided by the Bank of Italy's Survey of Household Income and Wealth (SHIW), reporting several socio-economic characteristics of Italian households for 2012. The SHIW is a biannual survey on the microeconomic behavior of Italian families, with a sample of approximately 8000 households per year. It contains information both on households (family composition) and on individuals. Moreover, it provides detailed information on several characteristics of the worker, such as net yearly earnings, average weekly hours of work and number of months of employment per year,[2] educational attainment (the highest completed school degree[3]), job experience, gender, marital status, sector of employment, composition of his/her family, parents background, regions of residence, and town size.

We consider adult heads of the family aged 19 or over, full time and part time employees, working either in the public or in the private sector[4] and such that information about earnings are available; this yields a total of 3141 individuals.

For each head of family we consider the following variables:

Y (Earnings): In its original formulation, the Mincer equation refers to the hourly price of labor as correct measure of worker's earnings.[5] SHIW contains annual earnings net of taxes and social security contributions, average number of hours worked per week, and number of months of employment per year. Based on these quantities, and as used by most empirical studies,[6] hourly wages are defined as

$$Y = \frac{\text{yearly net earnings}}{\text{months worked } \times \text{ weekly hours worked } \times 4}.$$

[2] Hourly wages are defined as: yearly net earnings/(months worked × weekly hours worked × 4).

[3] Standard and not actual year of formal schooling are recorded. Since students who fail to reach a standard have to repeat the year, the actual number of years is likely to be underestimated.

[4] We exclude self-employed because of the low reliability of their declared earnings.

[5] Monthly or annual wages would in addition capture the effect of individual's decisions on working hours. Given the only weak positive correlation between working time and educational attainment, it is reasonable to assume that the choice of hours worked reflects individual preferences rather than educational levels.

[6] Notice that hourly measure of earnings can be affected by measurement errors due to the fact that we calculate hourly wages as total earnings divided by hours of work; for instance, there might be part-time workers that do 2 weeks a month committing the whole day.

X_1 (Education): Education is generally measured by the number of years spent at school. SHIW does not contain information about this number, but only on the highest degree attained by individual. Following a common approach in literature [10, 54], we calculate the educational attainment of the individual by imputing the number of years required to complete her/his reported level of educational attainment.[7] More precisely, we consider that the (statutory) numbers of years required to obtain a primary and a junior school certificate is 5 and 8 years respectively; instead, for the upper secondary school the number of years ranges from 11 (vocational or technical school) to 13 (classical or scientific studies); finally, for tertiary education, we consider 16, 18, and 21 years for the university diploma, the college degree, and the postgraduate degree, respectively.

X_2 (Experience): Many empirical studies use age as a proxy for the (working) experience of individuals. But this choice can be severely biased, especially for young cohorts. Other authors use potential experience, defined as the difference between the current age and the age at the labor market entry, but they ignore the possibility of unemployment or underemployment, again a crucial feature for young cohorts. Here we use as proxy for experience the number of years for which a worker has been paid social security contribution; they should reflect the effective years of training on the job and learning-by-doing activities.

W_1 (Gender): gender, a dichotomous variable assuming values "Male" and "Female." The former is chosen as reference category.

W_2 (Area): area, a nominal variable with values "North," "Center," and "South." The former is chosen as reference category.

W_3 (Citizenship): citizenship, a dichotomous variable assuming values "Italian" and "Foreign." The former is chosen as reference category.

Previous studies have shown that native workers are likely to receive higher returns than immigrants [23], males to receive higher returns than female, and workers in the North of Italy to receive higher returns than workers in the other regions [6, 14]. We wanted our model to incorporate this knowledge, so we selected W_1 (Gender), W_2 (Area), and W_3 (Citizenship) as concomitant variables. Their role is, by means of the first term in (5), to drive individuals in the latent subpopulation in which individuals with similar socio-demographic status belong to.

Model (2) is fitted to the data for values of k ranging from 1 to 10; the model corresponding to the lowest BIC value has $k = 3$ components, and the corresponding estimated parameters are reported in Table 1. As the dependent

[7] Standard, not actual, years of formal schooling are recorded. Since students who fail to reach a standard have to repeat the year, the actual number of years is likely to be underestimated.

Table 1 Parameters estimates for model (2) with $k = 3$

		$j = 1$	$j = 2$	$j = 3$
Covariates	β_{j0} (Intercept)	**1.49314**	**1.37273**	**1.56716**
	β_{j1} (Education)	**0.03844**	**0.06528**	**0.02752**
	β_{j2} (Experience)	0.01180	**0.01670**	**0.02161**
	β_{j3} (Experience^2)	0.00034	−0.00012	**−0.00025**
	σ_j	0.75445	0.27891	0.19186
Concomitant variables	α_{j0} (Intercept)	0.00000	**3.01686**	**2.58453**
	α_{j1} (Gender: Female ref. Male)	0.00000	**−1.37438**	**−0.50159**
	α_{j2} (Area: Center ref. North)	0.00000	−0.37758	−0.36252
	α_{j3} (Area: South ref. North)	0.00000	**−1.72404**	**−1.35943**
	α_{j4} (Citizenship: Foreign ref. Italian)	0.00000	**−4.07112**	**−1.31669**
Relative size of the groups		0.05285	0.41006	0.53709

Bold style highlights regression coefficients significantly different from zero (significance level equal to 0.05)

Table 2 Conditional frequencies of group membership given the categories of the concomitant variables

	Gender		Area			Citizenship		Group
Group	Male	Female	North	Center	South	Italian	Foreign	relative size
1	0.03946	0.08440	0.03556	0.04444	0.08078	0.03841	0.18812	0.05285
2	0.50159	0.19444	0.44421	0.47302	0.32776	0.45349	0.00330	0.41006
3	0.45896	0.72115	0.52022	0.48254	0.59146	0.50810	0.80858	0.53709
Total	1.00000	1.00000	1.00000	1.00000	1.00000	1.00000	1.00000	1.00000

variable is the log of earnings, the estimation results show that, on average, one additional year of education increases earnings by around 3.844% for individuals in the first group, 6.528% for individuals in the second group, and 2.752% for individuals in the third group. This means that 41% only of the workers receive a good return from education's investment. Note that the average returns found by Brunello et al. [11] are 4–5%, so that we may see these previous results as a weighted average of our clusters' estimations.

The concomitant variables in the mixture model allow us to characterize the profile of the groups. As an example, the negative and significant coefficient associated to the Female category of the Gender variable for the second group tells us that being a woman makes less likely to belong to the second group than to the first (more precisely, its logit decreases by -1.37438).

Using the MAP operator, we can assign each subject to a group. Table 2 reports the conditional frequencies of group membership given each of the categories of the concomitant variables, while Table 3 reports the conditional frequencies of the categories of the concomitant variables given the groups.

The second group, which is the one with the highest return to education, is characterized by a disproportionately high presence of males; in fact, whereas 50.159% of all men belong to the second group, only 19.444% of women do

Table 3 Conditional frequencies of the categories of the concomitant variables given the groups

	Gender		Area			Citizenship	
Group	Male	Female	North	Center	South	Italian	Foreign
1	0.52410	0.47590	0.30723	0.16867	0.52410	0.65663	0.34337
2	0.85870	0.14130	0.49457	0.23137	0.27407	0.99922	0.00078
3	0.59988	0.40012	0.44221	0.18020	0.37759	0.85477	0.14523
Sample	0.70201	0.29799	0.45654	0.20057	0.34288	0.90353	0.09646

(Table 2). On the other hand, women are overrepresented in the third group, the one with the lowest return to education, where they account for 72.115%, whereas men are only 45.896% (Table 2). These gender differences are consistent with the findings of [18, 19]. Apart from discriminatory attitudes and specifically employers' unwillingness to invest in training female workers, gender differences have been ascribed to the different labor participation pattern between men and women arising from an intermittent female labor supply that may erode job skills [36]. Also, as pinpointed by Mincer [35], the decline in mortality that initiated the demographic transition increased incentives in investing in human capital, because the rise in longevity implied a higher profitability of investing in children's education [30, 31, 40]. At a later stage of the demographic transition, the costs of raising children increased with the increase in the cost of time, producing "an apparent trading of numbers of children for their quality" [35]. Note that mothers traditionally have a major role in raising children, so the opportunity cost for child care is greater for more educated women, especially where public daycare services are less developed as it is the case of Southern Italy [49].

Foreign workers, who embody 9.645% of the sample, are practically not present in the second group, whereas they account for 14.523% of the third group (Table 3). Evidence for a lower return on education for foreign immigrants in Italy is found in [1]. Using data on immigrants in Israel, [20] shows how once in the new country, immigrants tend to accept any kind of job, even ones in which they cannot fully exploit their human capital and skills. Workers in the South of Italy (34.288% of the sample) appear underrepresented in the second group (27.407%), while they account for more than a half of the first group (52.41%; see Table 3). Returns to education, so, appear slightly lower in the South of Italy; this is consistent with the findings of [18].

5 Conclusions

The Mincer human capital earnings function provides the most popular indicator of return to education. However, empirical studies have shown that populations of interest are often constituted by latent groups bearing different returns to educations and characterized by different socio-demographic profiles. In this paper, we propose a new mixture-based approach to make the Mincer earnings function more flexible with respect to unobserved heterogeneity. The proposed model allows to estimate the density of the income distribution, to detect homogeneous subpopulations, and to analyze the position of individuals with specific characteristics. The method is illustrated using data provided by the Bank of Italy's Survey of Household Income and Wealth in 2012. Our empirical results demonstrate that this method can be successfully used in practice.

Note that within specific applications, the Mincer function has been extended in different ways, notably to deal with the potential endogeneity of schooling [53], non-linear education premiums [21], heterogeneity in human capital within educational levels or work experiences [37], and cohort effects [23]. We acknowledge the relevance of these issues, although here, to keep our exposition simple, we focused on the classical Mincer equation. However, when required from the context of the application, these extensions can be easily incorporated within the framework proposed.

References

1. Accetturo, A., Infante, L.: Immigrant earnings in the italian labour market. Bank of Italy Temi di Discussione (Working Paper) No 695 (2008)
2. Amarante, V.: Income inequality in Latin America: data challenges and availability. Soc. Indic. Res. **119**(3), 1467–1483 (2014)
3. Atkinson, A.B.: The Economics of Inequality. Cambridge University Press, Cambridge (1986)
4. Bagnato, L., Punzo, A.: Finite mixtures of unimodal beta and gamma densities and the k-bumps algorithm. Comput. Stat. **28**(4), 1571–1597 (2013)
5. Bagnato, L., Punzo, A., Zoia, M.G.: The multivariate leptokurtic-normal distribution and its application in model-based clustering. Can. J. Stat. **45**(1), 95–119 (2017)
6. Battisti, M.: Reassessing segmentation in the labour market: an application for Italy 1995–2004. B. Econ. Res. **65**(s1), s38–s55 (2013)
7. Berta, P., Ingrassia, S., Punzo, A., Vittadini, G.: Multilevel cluster-weighted models for the evaluation of hospitals. METRON **74**(3), 275–292 (2016)
8. Björklund, A., Kjellström, C.: Estimating the return to investments in education: how useful is the standard Mincer equation? Econ. Educ. Rev. **21**(3), 195–210 (2002)
9. Boeri, T., Garibaldi, P.: Two tier reforms of employment protection: a honeymoon effect? Econ. J. **117**(521), F357–F385 (2007)
10. Brunello, G., Miniaci, R.: The economic returns to schooling for italian men. An evaluation based on instrumental variables. Labour Econ. **6**(4), 509–519 (1999)
11. Brunello, G., Comi, S., Lucifora, C.: The returns to education in Italy: a new look at the evidence. In: Harmon, C., Walker, I., Nielsen, N.W. (eds.) The Returns to Education in Europe. Edward Elgar, Cheltenham (2001)
12. Card, D.: The causal effect of education on earnings. Handbook of Labor Economics, vol. 3, pp. 1801–1863 (1999)
13. Card, D.: Estimating the return to schooling: progress on some persistent econometric problems. Econometrica **69**(5), 1127–1160 (2001)
14. Cipollone, P.: Is the Italian labour market segmented? Tech. rep., Bank of Italy, Economic Research and International Relations Area (2001)
15. Dang, U.J., Punzo, A., McNicholas, P.D., Ingrassia, S., Browne, R.P.: Multivariate response and parsimony for Gaussian cluster-weighted models. J. Classif. **34**(1), 4–34 (2017)

16. Dayton, C.M., Macready, G.B.: Concomitant-variable latent-class models. J. Am. Stat. Assoc. **83**(401), 173–178 (1988)
17. Dempster, A., Laird, N., Rubin, D.: Maximum likelihood from incomplete data via the EM algorithm. J. R. Stat. Soc. Ser. B Methodol. **39**(1), 1–38 (1977)
18. Fiaschi, D., Gabbriellini, C.: Wage functions and rates of return to education in Italy. In: Fifth Meeting of the Society for the Study of Economic Inequality (ECINEQ) Bari (Italy) (2013)
19. Flabbi, L.: Returns to schooling in Italy, OLS, IV and gender differences. In: Working Paper: Serie di Econometria ed Economia Applicata, Università Bocconi (1999)
20. Friedberg, R.M.: You can't take it with you? Immigrant assimilation and the portability of human capital. J. Labor Econ. **18**(2), 221–251 (2000)
21. Ghosh, P.K.: The contribution of human capital variables to changes in the wage distribution function. Labour Econ. **28**, 58–69 (2014)
22. Grün, B., Leisch, F.: **FlexMix** version 2: finite mixtures with concomitant variables and varying and constant parameters. J. Stat. Softw. **28**(4), 1–35 (2008)
23. Henderson, D.J., Polachek, S.W., Wang, L.: Heterogeneity in schooling rates of return. Econ. Educ. Rev. **30**(6), 1202–1214 (2011)
24. Ingrassia, S., Punzo, A.: Decision boundaries for mixtures of regressions. J. Korean Stat. Soc. **45**(2), 295–306 (2016)
25. Ingrassia, S., Minotti, S.C., Punzo, A.: Model-based clustering via linear cluster-weighted models. Comput. Stat. Data Anal. **71**, 159–182 (2014)
26. Ingrassia, S., Punzo, A., Vittadini, G., Minotti, S.C.: The generalized linear mixed cluster-weighted model. J. Classif. **32**(1), 85–113 (2015)
27. Karlis, D., Xekalaki, E.: Choosing initial values for the EM algorithm for finite mixtures. Comput. Stat. Data Anal. **41**(3–4), 577–590 (2003)
28. Lemieux, T.: The "Mincer equation" thirty years after schooling, experience, and earnings. In: Grossbard, S. (ed.) Jacob Mincer: A Pioneer of Modern Labor Economics, pp. 127–145. Springer, New York (2006)
29. Maruotti, A., Punzo, A.: Model-based time-varying clustering of multivariate longitudinal data with covariates and outliers. Comput. Stat. Data Anal. **113**, 475–496 (2017)
30. Mazza, A., Punzo, A.: Graduation by adaptive discrete beta kernels. In: Giusti, A., Ritter, G., Vichi, M. (eds.) Classification and Data Mining, Studies in Classification, Data Analysis and Knowledge Organization, pp. 243–250. Springer, Berlin (2013)
31. Mazza, A., Punzo, A.: **DBKGrad**: an R package for mortality rates graduation by discrete beta kernel techniques. J. Stat. Softw. **57**(Code Snippet 2), 1–18 (2014)
32. Mazza, A., Punzo, A., Ingrassia, S.: **flexCWM**: a flexible framework for cluster-weighted models. J. Stat. Softw. **86**(2), 1–30 (2018)
33. McLachlan, G.J., Basford, K.E.: Mixture models: inference and applications to clustering. Statistics Series, vol. 84. Marcel Dekker, New York (1988)
34. Mincer, J.: Schooling, Experience, and Earnings. National Bureau of Economic Research, New York (1974)
35. Mincer, J.: Technology and the labor market. In: Grossbard, S., Mincer, J. (eds.) A Pioneer of Modern Labor Economics, pp. 53–77. Springer, Boston (2006)
36. Mincer, J., Polachek, S.: Family investments in human capital: earnings of women. J. Polit. Econ. **82**(2, Part 2), S76–S108 (1974)
37. Nieto, S., Ramos, R.: Overeducation, skills and wage penalty: evidence for spain using piaac data. Soc. Indic. Res. **134**(1), 219–236 (2016)
38. Oreopoulos, P., Petronijevic, U.: Making college worth it: a review of research on the returns to higher education. Tech. rep., National Bureau of Economic Research (2013)
39. Patrinos, H.A.: Estimating the return to schooling using the Mincer equation. IZA World of Labor **278**, 1–11 (2016)
40. Punzo, A.: Discrete beta-type models. In: Locarek-Junge, H., Weihs, C. (eds.) Classification as a Tool for Research, Studies in Classification, Data Analysis and Knowledge Organization, pp. 253–261. Springer, Berlin (2010)

41. Punzo, A.: Flexible mixture modeling with the polynomial Gaussian cluster-weighted model. Stat. Model. **14**(3), 257–291 (2014)
42. Punzo, A., Ingrassia, S.: Parsimonious generalized linear Gaussian cluster-weighted models. In: Morlini, I., Minerva, T., Vichi, M. (eds.) Advances in Statistical Models for Data Analysis, Studies in Classification, Data Analysis and Knowledge Organization, pp. 201–209. Springer, Basel (2015)
43. Punzo, A., Ingrassia, S.: Clustering bivariate mixed-type data via the cluster-weighted model. Comput. Stat. **31**(3), 989–1013 (2016)
44. Punzo, A., McNicholas, P.D.: Robust clustering in regression analysis via the contaminated Gaussian cluster-weighted model. J. Classif. **34**(2), 249–293 (2017)
45. Punzo, A., Browne, R.P., McNicholas, P.D.: Hypothesis testing for mixture model selection. J. Stat. Comput. Simul. **86**(14), 2797–2818 (2016)
46. Punzo, A., Ingrassia, S., Maruotti, A.: Multivariate generalized hidden Markov regression models with random covariates: physical exercise in an elderly population. Stat. Med. **37**(19), 2797–2808 (2018)
47. Punzo, A., Mazza, A., Maruotti, A.: Fitting insurance and economic data with outliers: A flexible approach based on finite mixtures of contaminated gamma distributions. J. Appl. Stat. **45**(14), 2563–2584 (2018)
48. Quandt, R.E., Ramsey, J.B.: Estimating mixtures of normal distributions and switching regressions. J. Am. Stat. Assoc. **73**(364), 730–738 (1978)
49. Rondinelli, C., Aassve, A., Billari, F.: Women's wages and childbearing decisions: evidence from Italy. Demogr. Res. **S12**(19), 549–578 (2010)
50. Smith, N., Westergaard-Nielsen, N.: Wage differentials due to gender. J. Popul. Econ. **1**(2), 115–130 (1988)
51. Subedi, S., Punzo, A., Ingrassia, S., McNicholas, P.D.: Clustering and classification via cluster-weighted factor analyzers. Adv. Data Anal. Classif. **7**(1), 5–40 (2013)
52. Subedi, S., Punzo, A., Ingrassia, S., McNicholas, P.D.: Cluster-weighted t-factor analyzers for robust model-based clustering and dimension reduction. Stat. Method Appl. **24**(4), 623–649 (2015)
53. Trostel, P., Walker, I., Woolley, P.: Estimates of the economic return to schooling for 28 countries. Labour Econ. **9**(1), 1–16 (2002)
54. Vieira, J.A.C.: Returns to education in Portugal. Labour Econ. **6**(4), 535–541 (1999)
55. Wedel, M., Kamakura, W.: Market Segmentation: Conceptual and Methodological Foundations, 2nd edn. Kluwer Academic Publishers, Boston (2000)
56. Zarei, S., Mohammadpour, A., Ingrassia, S., Punzo, A.: On the use of the sub-Gaussian α-stable distribution in the cluster-weighted model. Iran. J. Sci. Technol. A **43**(3), 1059–1069 (2019)

Changes in Couples' Bread-Winning Patterns and Wife's Economic Role in Japan from 1985 to 2015

Miki Nakai

Abstract The trend towards dual-income families can be detected in recent years in many industrialized countries. However, despite the continuing rise in Japanese women's rates of participation in the economy over the period of industrialization and beyond, the notion of gendered division of labour has been seen as "normal" in Japanese society. The aim of this paper is to examine whether the determinants of married women's labour force participation have changed over the past several decades. Based upon social survey of national sample in Japan conducted in 1985, 1995, 2005, and 2015, we analyse the income provision-role type of the dual-income couples and examine change/stability of the factors that differentiate couples where the husband provides the majority of the couple's income from equal providers. We find the changing effects of women's own human capital on contribution to household income. On the other hand, the division of labour within households has not changed a lot over the past several decades.

Keywords Gender division of labour · Male breadwinner · Wives' economic dependency

1 Introduction

1.1 Background

A clear division of paid and unpaid work along gender lines is found in every country of the world, but the trend towards dual-income families can be detected in recent years in many advanced industrial societies.

M. Nakai (✉)
Department of Social Sciences, College of Social Sciences, Ritsumeikan University, Kyoto, Japan
e-mail: mnakai@ss.ritsumei.ac.jp

F. Greselin et al. (eds.), *Statistical Learning of Complex Data*,
Studies in Classification, Data Analysis, and Knowledge Organization,
https://doi.org/10.1007/978-3-030-21140-0_14

However, despite the continuing rise in Japanese women's participation in the economy as well as many Western societies, gender division of labour has been accepted as "normal" and still strong. While the number of households with wives entirely dependent on their spouses' income has dramatically declined, most women in dual-income couples still earn much less than their spouses, and households in which wives earn equal to or more than their husbands have been very few.

As gender inequalities in the division of labour at home are closely related to gender inequalities in other spheres of life, particularly in the labour market, understanding what determine the division of labour within Japanese couples is key to understanding other aspects of gender stratification. Many studies have argued that women's economic dependency on men is an important attribute of stratification systems and essential force in the maintenance of gender inequality (e.g. [12]).

The aim of the present study is to examine how couples' bread-winning patterns such as male-breadwinner couple, equal-provider couple, or female-breadwinner couples, relate to individual characteristics and how these associations have changed over time in Japan. Dual-income couples might not necessarily mean liberate women from their traditional gender role. There might be quite a gap between the households that wife's employment is perceived as secondary to her husband's and other households that have more symmetrical roles, or a more balanced sharing of responsibilities within the marriage. Therefore, the analysis places emphasis on what differentiates equal-provider couples from male-breadwinner couples among dual-income couples. In the following section, we describe several hypotheses related to couples' bread-winning patterns. To examine those hypotheses we perform multinomial logistic regression on the survey data collected in Japan. In Sect. 2, we describe the data and the variables of interest. In Sect. 3, we present the results of our analysis. In Sect. 4, we conclude the paper.

1.2 Hypotheses

Based on some previous studies, hypotheses are as follows.

Human Resource Hypothesis Women's improved educational opportunities are thought to boost female labour force participation in many countries. Also, it is considered that more females of the recent cohorts enter the labour market than those of older cohorts due to expanding access to higher education. Therefore, we first hypothesize that women's education may have positive effects on women's share of household income and therefore being equal provider.

However, the effect of a woman's educational attainment on her employment has not been significant in Japan (e.g. [2]). Our previous study also supported the notion that women are highly educated but typically barred from making full use of their education in economic and political fields up to the present [7, 8]. Having said that, woman's human resources might positively be associated with an increased likelihood that she is an equal provider relative to a secondary provider, once she overcomes the first hurdle, or resignation due to marriage or child birth.

Supplement Household Income Hypothesis Secondly, husband's socio-economic status may have negative effects on married women to become equal provider (e.g. [13]). Married women may be more likely to enter the labour market when the husband's income is low in order to supplement household income. According to past empirical research, women were employed in paid work for economic necessity, and it is not until the 1970s that women started to pursue careers [4].

However, there has been significant diversity in the impact of husbands' resources on their spouses' employment since around the end of the twentieth century and distinct differences in the impact of husbands' resources on their spouses' employment behaviour correspond to the welfare state regimes (e.g. [1, 3, 10, 14]). For example, in conservative continental European welfare states such as Italy, France and Germany, the association is negative as it used to be; for men with high occupational resources to suppress spouse's participation in paid work, showing the traditional division of labour in couples and increasing dependency of married women on their spouses over the life course. In social democratic welfare states, on the other hand, male's occupational resources increase women's labour market activity (positive association). Positive effect implies that economic resource at the household level facilitates a woman's employment also because it helps balancing work and family. More and more advanced postindustrial economies see the positive effects of husband's occupational resources on their partner's participation rates in recent years. Women married to well-educated husbands as well as women with high-income partners are less likely to leave the labour market than women with low-resource partners.

Values Hypothesis Thirdly, we also hypothesize that values and attitudes toward the family and gender roles may affect couple's bread-winning patterns. We hypothesize that gender egalitarian attitudes are positively associated with the probability of being in an equal-provider couple. For example, given that younger cohorts are more egalitarian than older cohorts, it may lead to the rise in equal-provider among younger couples. Inglehart and Norris [5] argue that the twentieth century gave rise to profound changes in traditional sex roles, but that the force of this "rising tide" has varied among rich and poor societies. They demonstrate that richer, postindustrial societies support the idea of gender equality more than agrarian and industrial societies and intergenerational differences in values are largest in postindustrial societies and relatively minor in agrarian societies, suggesting that the former are undergoing intergenerational changes in values. They also argue that cohort change in gender-role attitudes in postindustrial societies is unidimensional, with newer cohorts consistently more egalitarian than older cohorts.

We also hypothesize that values related to the household context may influence gendered arrangement for work and care in the household. We focus on the degree of educational homogamy.[1] Whether or not the couple is homogamous seems to be associated with a patriarchal culture. More patriarchal households, which may be associated with female hypergamous couples, may prefer traditional marriage practice; women are expected to fulfil the roles of wife and mother, and men are expected to be the chief provider. These asymmetric gender relation within the marriage may influence couples' preferences for bread-winning type, as well as their relative power within the marriage [11].

Lifestage Restriction Hypothesis The burden of child-rearing poses a formidable obstacle to women's professional ambitions and have women accept a secondary provider role within household. Even though gender equality matters in many societies, most research found that women have had primary responsibility for household chores, as well as caring for their children. Our previous research also showed that the presence of preschool children has strong negative influence on wife's labour participation.

2 Data and Methods

Data for the present study were obtained from the past three decades of four waves of cross-sectional data: the 1985, 1995, and 2005 Social Stratification and Social Mobility (SSM) surveys of Japanese society, and the 2015 Stratification and Social Psychology (SSP) survey in Japan. All the surveys were conducted with similar approach: face-to-face interviews with a special focus on social stratification and inequality in contemporary Japan. All the surveys selected national representative respondents through multiple-stage sampling. The subjects of these surveys were men and women, aged between 20 and 69 for the surveys in 1985, 1995, and 2005, and between 20 and 64 for the 2015 SSP survey. Data were collected from 1248 men and 1405 women in 1985, 2490 men and 2867 women in 1995, 2660 men and 3082 women in 2005, and 1644 men and 1931 women in 2015. The response rates were 67.9%, 66.0%, 44.1%, and 43.0% in 1985, 1995, 2005, and 2015, respectively.

To make data comparable across the four datasets, we limit our analysis to the householders and their spouses, where wife's age is between 25 and 54. The available data refer to 9067 respondents (994 in 1985, 3180 in 1995, 2862 in 2005, and 2031 in 2015). Using multinomial logistic regressions we analyse how individual- and household-level characteristics are associated with each of the three dual-income types. We estimate the effects of the correlates on the odds of being equal-provider or female-breadwinner couples (reference category is male-breadwinner couples) in each of the four waves.

[1]Homogamy is defined as marriage of both the husband and wife having similar levels of educational attainment. Hypergamy is when the wife is less educated than the husband, and hypogamy is where the wife is more educated than the husband.

Dependent Variable We focus on within-couple inequality in the household. We use a concept of wives' contribution to household income as an aspect which reflects within-couple inequality in the household, which is defined (a) income provision-role type, and (b) wives' contribution to total household income. In the present study, we analyse (a) income provision-role type as a dependent variable.

Income provision-role type is measured based on whether a dominant provider exists and identifies who she/he may be. We use a five group classification: (1) husband sole provider, (2) husband provides majority, (3) equal providers, (4) wife provides majority, (5) wife sole provider [9]. Although we first show the distribution of five-category household type in Table 1, we restrict the subsequent analysis to the couples of the three dual-income groups (2, 3, and 4) to estimate a multinomial logistic model for examining the factors associated with the equal-provider couples and female-breadwinner couples as opposed to male-breadwinner couples. This restriction reduced sample to 4024 (420 in 1985, 1465 in 1995, 1112 in 2005, and 1027 in 2015).

Independent Variables To capture the effects of human resources of women, we include wife's education. Wife's education is collapsed into four categories: (1) less than a high school, (2) high school, (3) 2-year college, and (4) 4-year tertiary education or more, where high school is the reference category. Wife's age is coded into six categories: (1) 25–29, (2) 30–34, (3) 35–39, (4) 40–44, (5) 45–49, and (6) 50–54, where 30–34 years is the reference category.

Married couples division of labour may vary systematically also with regards to household-level characteristics. The household level explanatory variables include age and the number of children within a household, husband's income, and the couples' relative education. The number of children has four categories: (1) no, (2) one, (3) two, and (4) three or more children, with 'no' as reference category. The presence of a preschooler is coded as a binary variable with respect to children's age 0–6, with no preschooler as reference category. Husband income level is measured by income decile (ten groups) in each survey year. The couples' relative education-level variable measures whether wife has higher or lower education than her spouse and has three categories: (1) husband and wife have equal education, (2) hypogamy, and (3) hypergamy, where equal educational level is the reference category.

3 Results

We first examined the division of paid and unpaid work between spouses within households. Table 1 shows how bread-winning patterns and average wife's economic contribution have changed over the past three decades. Wife's contribution to household income, which is the percentage of income contributed by wives, was calculated by respondent's and spouse's annual incomes. Although an overwhelming majority (70%) of couples were dual-income by the year 2015, most of them

Table 1 Trends in percent distribution of household types of couples and wife's economic contribution to household income: 1985–2015

	1985	1995	2005	2015
Household type				
Husband sole provider	42.8%	42.0%	41.3%	29.0%
Husband provides majority	46.8%	47.7%	44.7%	51.4%
Equal providers	8.9%	8.7%	11.2%	14.1%
Wife provides majority	1.6%	1.2%	1.9%	5.2%
Wife sole provider	0.0%	0.4%	0.9%	0.3%
Wife's economic contribution				
All age	14.0	15.1	18.6	25.6
Wife aged between 25–54	15.1	14.9	17.8	23.1

are male-breadwinner couples, which is often be considered to be associated with low gender egalitarian attitudes. The table also shows that equal-provider couples are only 14% in Japan even in 2015.

We present the results of multinomial logistic regression in Table 2.[2]

First, our hypothesis that wife's own education would be positively associated with the probability of being in an equal-provider couples among dual-income couples was supported. Having a college education heightened a wife's likelihood that she was an equal provider relative to a secondary provider, compared to women with a medium level of education, on the condition that the married couple households with college educated wives are dual-income since the mid-1990s. Although tertiary education has not been positively associated with women's participation in paid work in Japan, when we focus on dual-income couples, women who have invested more in their own human capital less readily settle for a secondary provider role than women who have invested less in their human capital accumulation since around the end of the twentieth century.

Second, the effects of husband's income have been remarkably significant and remain constant over time. Women who have husbands with low annual income are more likely to report being in an equal-provider couple, as opposed to a male-breadwinner couple. This suggests that women's participation in paid work in Japan is primarily driven by economic necessity, in fact, the purpose of keeping the level of household income, rather than pursuing careers.

Third, we do not find consistent significant association between age and the probability of being in an equal-provider versus a male-breadwinner couple. We expected a degree of gender egalitarian values would be reflected in gender equality in couples' earnings structures. However, this was not supported by our findings and it is still not normative for young married women to share equally in providing

[2]Because our primary interest is to understand what factors differentiate equal-provider from male-breadwinner couples, the part with regards to the estimated coefficients affecting the probability to belong to female-breadwinner instead of male-breadwinner couples is not shown in Table 2.

Table 2 Multinomial logistic regression estimates: 1985–2015

		Equal vs. male breadwinner			
		1985	1995	2005	2015
Age (ref: 30–34)	25–29	0.192	−0.151	−0.826**	0.064
		(0.574)	(0.363)	(0.382)	(0.319)
	35–39	0.027	−0.046	−0.134	0.017
		(0.517)	(0.288)	(0.289)	(0.294)
	40–44	0.516	−0.293	−0.180	−0.009
		(0.591)	(0.317)	(0.306)	(0.322)
	45–49	0.094	0.285	−0.113	0.191
		(0.628)	(0.311)	(0.334)	(0.332)
	50–54	0.324	0.118	−0.048	0.715**
		(0.621)	(0.331)	(0.336)	(0.344)
Wife's education	< high school	−1.086***	−0.138	−1.845**	−0.627
(ref: high school)		(0.417)	(0.253)	(0.741)	(0.786)
	Two-year college	−0.068	0.181**	0.592**	0.572**
		(0.487)	(0.235)	(0.258)	(0.225)
	Four-year college	0.143	1.348***	1.762***	1.448***
		(0.609)	(0.239)	(0.238)	(0.315)
Couple's education	Husband > wife	0.484	0.178	−0.157	0.336
(ref: equal)		(0.368)	(0.205)	(0.239)	(0.204)
	Husband < wife	0.493	−0.156	−0.062	0.029
		(0.397)	(0.229)	(0.258)	(0.256)
Husband's income		−0.229***	−0.105***	−0.261***	−0.204***
		(0.062)	(0.032)	(0.040)	(0.036)
Number of children	1	−0.928	−0.761**	−0.475	0.032
(ref: 0)		(0.659)	(0.344)	(0.310)	(0.300)
	2	−1.152*	−0.781***	−0.905***	−0.288
		(0.612)	(0.298)	(0.278)	(0.280)
	3 or more	−1.164*	−1.237***	−1.262***	−0.304
		(0.663)	(0.332)	(0.319)	(0.326)
Preschool children	Yes	−0.119	0.450*	0.189	0.467*
(ref: no)		(0.490)	(0.253)	(0.266)	(0.248)
Constant		0.685	−0.829	0.793	−0.596
		(0.733)	(0.327)	(0.349)	(0.352)
Nagelkerke (Pseudo) R^2		0.241	0.112	0.291	0.309

Standard Errors are in parentheses below the estimates. * $p<0.10$, ** $p<0.05$, *** $p<0.01$

income. Younger couples may also face challenges in work-family reconciliation as well as older couples.

Finally, the probability of being in an equal-provider couple decreases with the number of children (not significant in 2015, though). This suggests that the more children a couple have, the greater likelihood they were in a couple that husband is a primary provider. However, interestingly, the presence of preschool children has no or very little effects on bread-winning arrangement, of which the sign and significance are not expected in our hypothesis. The previous study found that the presence of preschool children strongly negatively affects wife's labour participation (e.g. [6]). However, this somewhat unpredictable positive effects might suggest polarization of occupational class-based outcomes among working mothers.

4 Conclusion and Discussion

We find the changing effects of women's own human capital on contribution to household income: education is important for women to increase the probability of having a more equal division of labour and time within the marriage rather than a unequal role allocation, but it is not until the late 1990s that highly educated women show a higher probability of belonging equal-provider couple rather than male-breadwinner couple. However, the division of labour in marriage has not changed a lot and wives' earnings still help to reduce income inequality across married couple households.

Analysing differences of values and availability of policy from comparative perspective in future research could enrich theory and evidence about how introduction of policy might affect employment of married women, especially mother of preschool children. Moreover, asymmetry in terms of what percentage of household income wife and husband provide may be correlated with other aspects of asymmetric relationship such as division of roles in the home and family or couple relationship, which we leave to future research.

Acknowledgements This work is supported by JSPS Grant-in-Aid for Scientific Research (No. 26380658, No. 17K04103), and (No. 16H02045) as part of the SSP Project. The author thanks both the Social Stratification and Social Mobility Committee for the permission to use the SSM data, and the SSP Project for the permission to use the SSP 2015 survey.

References

1. Blossfeld, H.P., Drobnič, S.: Careers of Couples in Contemporary Societies. From Male Breadwinner to Dual Earner Families. Oxford University Press, Oxford (2001)
2. Brinton, M.C.: Women and the Economic Miracle: Gender and Work in Postwar Japan. University of California Press, Berkeley (1993)
3. Esping-Andersen, G.: Social Foundations of Postindustrial Economies. Oxford University Press, Oxford (1999)
4. Goldin, C.: The quiet revolution that transformed women's employment, education and family. Am. Econ. Rev. **96**(2), 1–21 (2006)
5. Inglehart, R., Norris, P.: Rising Tide: Gender Equality and Cultural Change Around the World. Cambridge University Press, Cambridge (2003)

6. Nagase, N.: Women's work choice: household production and labour supply. In: Chuma, H., Suruga, T. (eds.) Changing Employment Practices and Femal Labor Force, pp. 279–312. University of Tokyo Press, Tokyo (1997)
7. Nakai, M.: Occupational segregation and opportunities for career advancement over the life course. Jpn. Sociol. Rev. **159**(4), 699–715 (2009)
8. Nakai, M.: Trends in women's career patterns and occupational mobility in Japan: analysis of the social stratification and mobility survey 1985–2005. Jpn. J. Res. Household Econ. **89**, 11–21 (2011)
9. Raley, S.B., Mattingly, M.J., Bianchi, S.M.: How dual are dual-income couples? Documenting change from 1970 to 2001. J. Marriage Fam. **68**(1), 11–28 (2006)
10. Sainsbury, D.: Gender and Welfare State Regimes. Oxford University Press, New York (1999)
11. Simpson, I.H., England, P.: Conjugal work roles and marital solidarity. In: Aldous, J. (ed.) Two Paycheck: Life in Dual-Earner Families, pp. 147–172. Sage, Beverly Hills (1982)
12. Sorensen, A., McLanahan, S.: Married women's economic dependency, 1940–1980. Am. J. Sociol. **93**, 659–687 (1987)
13. Treas, J.: The effect of women's labor force participation on the distribution of income in the United States. Annu. Rev. Sociol. **13**, 259–288 (1987)
14. Vitali, A., Arpino, B.: Who brings home the bacon? The influence of context on partners' contributions to the household income. Demogr. Res. **35**, 1213–1244 (2016)

Weighted Optimization with Thresholding for Complete-Case Analysis

Graziano Vernizzi and Miki Nakai

Abstract Complete-case analysis, also known as listwise deletion method (LD), is a relatively popular technique to handle datasets with incomplete entries. It is known to be effective when data are missing completely at random. However, by reducing the size of the dataset it can weaken the final statistical analysis. We present an optimization algorithm that improves the size of the final dataset after applying LD. It is based on a constrained weighted optimization technique to determine the maximum number of variables and respondents from the initial dataset that are preserved after applying LD. The main feature is that the method allows for selecting a specific set of variables (or respondents) that must be kept during the optimization, while balancing their relative importance by means of suitable weights. Moreover, we provide analytic formulas for the optimal solution, that can be easily evaluated numerically, reducing the computational complexity associated to the usage of off-the-shelf packages for solving similar large constrained optimization problems. We illustrate the application of our weighted optimization method to some examples and real datasets.

Keywords Missing data · Complete-case analysis · Constrained optimization

1 Introduction

Datasets are often plagued by incomplete entries, due to a variety of reasons: improper codes, faulty recording, missed questions, to name a few. Most statistical analysis procedures cannot incorporate missing data directly, and therefore a

G. Vernizzi
Department of Physics and Astronomy, Siena College, Loudonville, NY, USA
e-mail: gvernizzi@siena.edu

M. Nakai (✉)
Department of Social Sciences, College of Social Sciences, Ritsumeikan University, Kyoto, Japan
e-mail: mnakai@ss.ritsumei.ac.jp

F. Greselin et al. (eds.), *Statistical Learning of Complex Data*,
Studies in Classification, Data Analysis, and Knowledge Organization,
https://doi.org/10.1007/978-3-030-21140-0_15

number of different methods have been developed to *eliminate* or *impute* all missing data before proceeding with any analysis, e.g. complete-case analysis, or multiple imputation and full-information maximum likelihood. Among the many options available nowadays, and that have been implemented in conventional software suites, an effective (albeit drastic) technique to handle incomplete datasets is the complete-case analysis, also known as listwise deletion (LD) method. According to LD, any observation with at least one missing data entry is removed completely from the dataset. It is evident that the LD is advantageous only when a small percentage of the data are excluded this way. However, in several practical situations, a brute-force application of LD can deplete the dataset to a point where the number of data entries is not sufficient for a meaningful subsequent statistical analysis. In this work, we show how one can improve the applicability of LD, by selecting a suitable subset of variables from the dataset. By introducing suitable weights, the selection algorithm allows for the inclusion of any subset of variables that are considered essential, and cannot be eliminated. For the sake of conciseness, we do not summarize here the broad literature discussing LD applicability, for which we refer to [1, 7, 11, 14]. We only mention that LD is known to work best when data are missing completely at random [2, 8, 10, 12].

2 Weighted Optimization

A dataset with L variables and N observations can be represented by a rectangular matrix X with N rows and L columns, with entries X_{ij} representing the value of the i-th observation for the j-th variable. The presence of missing data can be recorded in a *shadow matrix* A [5]: where entries are $A_{ij} = 0$ (complete value) or $A_{ij} = 1$ (missing value). In situations where missing values abound, LD can reduce the size of the statistical sample dramatically. In such cases, it may be advantageous to exclude combinations of variables that are particularly plagued by missing values. However, *which* rows and columns should one delete in order to *maximize* the number of entries that remain, after applying LD? We illustrate the problem with an example: given the shadow matrix A for a dataset with $N = 4$ observations and $L = 3$, there are different combinations of variables that can be removed before applying LD:

$$A = \begin{pmatrix} 0 & 0 & 0 \\ 0 & 0 & 1 \\ 0 & 1 & 0 \\ 0 & 0 & 1 \end{pmatrix} \quad \Rightarrow \quad A = \begin{pmatrix} 0 & 0 & 0 \\ \not{0} & \not{0} & \not{1} \\ \not{0} & \not{1} & \not{0} \\ \not{0} & \not{0} & \not{1} \end{pmatrix}; \quad A = \begin{pmatrix} 0 & \not{0} & \not{0} \\ 0 & \not{0} & \not{1} \\ 0 & \not{1} & \not{0} \\ 0 & \not{0} & \not{1} \end{pmatrix}; \quad A = \begin{pmatrix} 0 & 0 & \not{0} \\ 0 & 0 & \not{1} \\ \not{0} & \not{1} & \not{0} \\ 0 & 0 & \not{1} \end{pmatrix}.$$

There are only three missing entries. However, a direct application of LD would delete all rows but the first one, wasting six non-missing values. A different possibility is to delete the last two variables, which saves four non-missing entries, but also loses five in the process. One can show that the best solution would be to remove the third variable only: LD leaves six non-missing entries, i.e. 50% of the original dataset. In general, there are combinations of variables (columns) in the matrix A that are optimal, in the sense that the number of remaining values after LD is maximal.

The problem can be formulated mathematically. We introduce two column vectors, r and c, whose elements r_i $(i = 1, \ldots, N)$ and c_j $(j = 1, \ldots, L)$ are binary numbers (0 or 1). The values r_i and c_j indicates whether the i-th row and j-column of X are deleted (zero) or not (one). For instance, the optimal solution in the last example corresponds to the vectors: $r = (1; 1; 0; 1)$ and $c = (1; 1; 0)$. For any given choice of r and c, the total number of missing entries $d(r, c)$ is: $d(r, c) = \sum_{i=1}^{N} \sum_{j=1}^{L} r_i c_j A_{ij} \equiv r^T Ac$, where r^T indicates the transposition operation, and all products are understood as matrix products throughout this article. The *total* number of missing entries in the dataset is $d_{total} = 1_N^T A 1_L$, where 1_x indicates the all-one column vector with x elements. The goal is to find what binary vectors r and c render $d(r, c) = 0$ (which effectively implements LD) and maximize the number of remaining non-missing entries, i.e. maximize $m(r, c) = r^T (1 - A)c$. Such a constrained optimization problem over the field of binary numbers $\{0, 1\}$ is a classic example of integer non-linear programming optimization, which is known to be NP-hard [4]. Much scientific literature transforms the discrete optimization problem into finding a global optimum over continuous variables [6]. However, the continuous version of $m(r, c)$ over all *real* vectors r, c is quadratic but not convex, and the global optimum is not guaranteed to exist [9]. Moreover, in several practical applications, there may be variables or respondents that one does not wish to see removed from the analysis. We therefore consider a different problem, which is to find what vectors r and c render $d(r, c) = 0$ and *maximize the number of variables and respondents that remain after LD*. By associating the weights $\omega_j \geq 1$ to each variable, and a weights $w_i \geq 1$ to each respondent, the problem we consider is the minimization of the (convex) quadratic *weighted functional*: $F_0 = (r - 1_N)^T W(r - 1_N) + (c - 1_L)^T \Omega(c - 1_L)$, where $\Omega = \delta_{ij}\omega_i$ and $W = \delta_{ij} w_i$ are a $L \times L$ and $N \times N$ diagonal matrices, respectively. The two problems are not independent. In general, the first problem maximizes the number of non-missing entries but does not guarantee to obtain the largest possible number of variables and respondents, not to mention the variables one wishes to keep during the optimization. The second problem maximizes the number of variables and respondents, but by doing so it may sacrifice some non-missing entries. For instance, in the example at the beginning of this section, the total number of variables and respondents preserved by the three cases are 4, 5, and 5, respectively, which is different from the number of non-missing entries 3, 4, and 6, respectively. The discrepancy can be mitigated in part since the second problem has several local minimizers in general, and in the Examples section we introduce a thresholding technique that can be used to select the minimizer with the highest number of non-missing entries. Surely, the main advantage of considering the second problem is that it is amenable to analytical treatment, and in fact it can be cast into an *unconstrained* optimization problem:

$$F(W, \Omega) = (r - 1_N)^T W(r - 1_N) + (c - 1_L)^T \Omega(c - 1_L) + 2\lambda d(r, c)\,, \qquad (1)$$

where we introduced the Lagrange multiplier 2λ for the constraint (the irrelevant factor 2 keeps the following expressions simple). Since W and Ω are positive definite matrices, so is also the quadratic form F_0. Moreover, the constrained minimization is on the closed set $d(r, c) = 0$, and a generalization of the Weierstrass theorem (see for instance [3]) guarantees the existence of a solution, which can be determined by the Lagrange method. Equation (1) can be minimized by determining the stationary points of F with respect to r, c, and λ:

$$\begin{cases} F'_r : \; W(r - 1_N) + \lambda A c = 0 \\ F'_c : \; \Omega(c - 1_L) + \lambda A^T r = 0 \\ F'_\lambda : \; r^T A c = 0 \end{cases} . \tag{2}$$

By solving the first equation with respect to r, the second equation with respect to c, and by substituting one into the other, we obtain:

$$\begin{cases} r = \left(\mathbb{I}_N - \lambda^2 W^{-1} A \Omega^{-1} A^T\right)^{-1} \left(1_N - \lambda W^{-1} A 1_L\right) \\ c = \left(\mathbb{I}_L - \lambda^2 \Omega^{-1} A^T W^{-1} A\right)^{-1} \left(1_L - \lambda \Omega^{-1} A^T 1_N\right) \end{cases} . \tag{3}$$

The matrices W^{-1} and Ω^{-1} always exist since they are positive definite. By inserting Eq. (3) in the third equation in (2), we obtain an equation for λ:

$$\begin{aligned} &\left(1_N^T - \lambda 1_L^T A^T W^{-1}\right) \left(\mathbb{I}_N - \lambda^2 A \Omega^{-1} A^T W^{-1}\right)^{-1} A \times \\ &\times \left(\mathbb{I}_L - \lambda^2 \Omega^{-1} A^T W^{-1} A\right)^{-1} \left(1_L - \lambda \Omega^{-1} A^T 1_N\right) = 0 . \end{aligned} \tag{4}$$

Such an equation simplifies considerably by using the singular value decomposition of the matrix $S \equiv W^{-1/2} A \Omega^{-1/2}$, which is $S = U^T \Sigma V$ where U is a $N \times N$ orthogonal matrix (i.e. $U^T U = U U^T = \mathbb{I}_N$), V is a $L \times L$ orthogonal matrix (i.e. $V^T V = V V^T = \mathbb{I}_L$), and Σ is a rectangular $N \times L$ matrix with diagonal elements only $\Sigma_{ij} = \delta_{ij} \sigma_i$. The singular values σ_i are the s positive eigenvalues of the matrix $S^T S$. By inserting $A = W^{1/2} U^T \Sigma V \Omega^{1/2}$ in Eq. (4), and using the matrix identity $(\mathbb{I} - B C B^{-1})^{-1} = B(\mathbb{I} - C)^{-1} B^{-1}$ repeatedly, the last equation reads:

$$\left(\rho^T - \lambda \gamma^T \Sigma^T\right) \left(\mathbb{I}_N - \lambda^2 \Sigma \Sigma^T\right)^{-1} \Sigma \left(\mathbb{I}_L - \lambda^2 \Sigma^T \Sigma\right)^{-1} \left(\gamma - \lambda \Sigma^T \rho\right) = 0 \tag{5}$$

where $\rho \equiv U W^{1/2} 1_N$, and $\gamma \equiv V \Omega^{1/2} 1_L$. Due to the particular structure of the matrix Σ, Eq. (5) can be written in terms of the singular values only:

$$\sum_{i=1}^{s} (\rho_i - \lambda \gamma_i \sigma_i)\, \sigma_i\, (\gamma_i - \lambda \sigma_i \rho_i) \,/\, \left(1 - \lambda^2 \sigma_i^2\right)^2 = 0 . \tag{6}$$

In general, Eq. (6) is a polynomial equation that does not admit a closed-form solution for λ, but it can be solved numerically.

There may be situations when the inverse matrices enclosed by parentheses in Eq. (3) do not exist, i.e. when the determinants $\det\left(\mathbb{I}_N - \lambda^2 W^{-1} A \Omega^{-1} A^T\right) = 0$ or $\det\left(\mathbb{I}_L - \lambda^2 \Omega^{-1} A^T W^{-1} A\right) = 0$. Such determinants are the characteristic equations for the eigenvalues of $\Sigma\Sigma^T$ and also $\Sigma^T\Sigma$, therefore, in terms of singular values that occurs only when $\lambda = 1/\sigma_i$ for some i. Equation (6) diverges at those values, which is an indication that the stationary points for $F(W, \Omega)$ in Eq. (1) are actually at $\lambda = \infty$. This fact can be implemented numerically by simply plugging in Eq. (3) a sufficiently large value for λ. An alternative approach for such cases is to multiply one of the matrices W or Ω by a constant factor: we found that even a small perturbation is sufficient to move the stationary points of Eq. (1) away from the singular values $\lambda = 1/\sigma_i$, which provides a finite solution for the optimization problem.

Finally, a word of caution: when N or L are large numbers, the numerical matrix inversion in Eq. (3) can be a computational daunting task. In such cases, we can approximate the inversion by a Neumann series: $\left(\mathbb{I} - \lambda^2 K\right)^{-1} = \sum_{i=0}^{\infty} \lambda^{2i} K^i$ (geometric series expansion). In addition, in the particular "weightless" limit with $\Omega = \mathbb{I}_L$ and $W = \mathbb{I}_N$, Eq. (3) simplify to:

$$r = \left(\mathbb{I}_N - \lambda^2 A A^T\right)^{-1} (1_N - \lambda A 1_L), \quad c = \left(\mathbb{I}_L - \lambda^2 A^T A\right)^{-1} \left(1_L - \lambda A^T 1_N\right). \tag{7}$$

Equation (6) still holds, with σ_i being the singular values of the shadow matrix A. We conclude this section by commenting on the fact that several algorithms are nowadays available for solving large constrained quadratic optimization problems numerically. Nevertheless, we believe that the analytic formulas Eqs. (3) and (6) provide a higher vantage point when optimizing the functional F in Eq. (1), since the numerical evaluation of the above analytic formulas is computationally simpler than solving the full initial combinatorial optimization problem.

3 Examples

We illustrate our proposed optimization method with three examples. First, by using the matrix A from the example at the beginning of Sect. 2, with all weights equal to 1, the singular value decomposition of A gives two singular values, $\sigma_1 = \sqrt{2}$, and $\sigma_2 = 1$, and the vectors $\rho = (\sqrt{2}; 1; 0; 1)$, $\gamma = (1; 1; 1)$. The polynomial equation (6) for the constraint is: $3 - 2\lambda - 10\lambda^2 + 2\lambda^3 + 8\lambda^4 = 0$. Such an equation admits two real solutions $\lambda \approx 0.92$ and $\lambda \approx 0.54$. After evaluating $F(W, \Omega)$ in Eq. (1) with those roots, we pick the latter value, which gives $r = (1, 1.1, 0.6, 1.1)$ and $c = (1, 0.6, -0.2)$. We tested several methods to represent the real values of r, c in terms of 0 and 1 entries. Among them we recommend using the following *thresholding method*. Define $c_i(t) \equiv \theta(c_i - t)$, where $\theta(\cdot)$ is the Heaviside step function and t the threshold with $\min c_i \leq t \leq \max c_i$. LD can be applied by setting

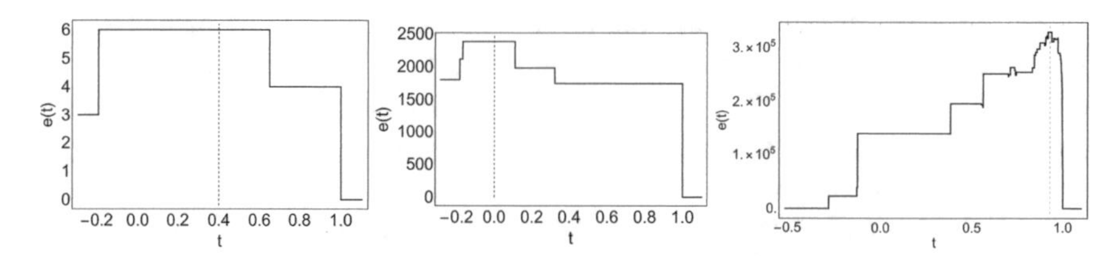

Fig. 1 Thresholding method applied to the example at the beginning of Sect. 2 (left), to the NLSY dataset (center), and to the SSP2015 dataset (right). The vertical axis $e(t)$ shows the number of non-missing entries obtained by the optimization method. The function $e(t)$ is maximal for a threshold value in a region around $t^* = 0.4$ (left), $t^* = 0.0$ (center), or $t^* = 0.9$ (right), respectively, see dashed lines. If $c_i < t^*$ its value is rounded to 0, otherwise it is rounded to 1

$r(t) = \delta_{0,Ac(t)}$. The fraction of non-missing entries that are still available after applying LD is $e(t) = r(t)(1_N 1_L^T - A)c(t)$. A simple plot of $e(t)$ in the interval $\min(c_i) \le t \le \max(c_i)$ shows a region with a maximum, whose location determines optimal threshold values for t. Figure 1(left) shows the plot for the example under consideration. For instance, by choosing the thresholding value $t = 0.4$, we obtain $c(t) = \{1; 1; 0\}$ and $r(t) = \{1; 1; 0; 1\}$. That leaves $e(t) = 6$ remaining non-missing entries, which is the optimal case as discussed at the beginning of Sect. 2.

As a second example, we apply our optimization method to the dataset used in chapter 4 of [2], where 581 children were interviewed in 1990 as part of the National Longitudinal Survey of Youth (NLSY). As in [2], we consider only the eight variables: "ANTI", "SELF", "POV", "BLACK", "HISPANIC", "DIVORCE", "GENDER", "MOMWORK" (see [2] for details). In this case, the complementary matrix A has rank $k = 4$, and gives a polynomial equation for λ of order fourteen. Four solutions are real, and among them only one corresponds to a minimum of F. The corresponding thresholding plot is in Fig. 1 (center). The thresholding value $t = 0.0$ gives $e(t) = 2382$ remaining non-missing entries, and $c = (1; 0; 0; 1; 1; 1; 1; 1)$, which means that if we discard the variables "SELF", and "POV" before applying LD, the maximum number of data entries is preserved. More precisely, the full dataset has 4038 not missing entries. If one applies LD directly (without optimization), only 1800 entries remain. However, our optimization algorithm finds that after removing the two variables "SELF" and "POV", LD leaves 2382 entries (32% increase). Interestingly, if one discards "POV" only (which is in fact the variable with most missing data, corresponding to 431 non-missing entries), LD leaves only 2114 entries (17% increase). To be absolutely certain, we performed an exhaustive combinatorial search over all $2^8 = 128$ possible binary vectors c, and verified the validity of this result.

As a third example, we test the possibility to add weights: the purpose is to either favor (or penalize) groups of variables that the user wishes to prioritize. For this example, we use the *2015 Japanese survey on Stratification and Social Psychology* (SSP2015) [13]. The survey collects face-to-face interviews of a randomly selected sample of the Japanese population represented by men and women with age 20–64. The dataset was compiled in 2015 from a total of 3575 respondents (1644 men and 1931 women) over 171 variables. However, 89752 entries are missing, corresponding to 14.7% of the whole dataset. When one applies LD directly on the whole dataset, *all* respondents get deleted. On the other hand, the optimized LD method with no weights (e.g. $W = \mathbb{I}_N$, $\Omega = \mathbb{I}_L$), produces a constraint equation for λ, Eq. (6), with only three real solutions, one of which corresponds to a minimum for F. By using the thresholding technique we find 114 variables that leave 2880 respondents. Such a solution is optimal, in the sense that no other group of variables leaves more data entries after applying LD. The solution found by the optimization algorithm in this example, corresponds to 53.7% of the initial dataset, which is a great improvement over the direct LD without optimization. Now, let us suppose we are interested in a specific subset of variables, that we consider particularly important and should not be deleted by the optimization algorithm. For instance, we pick the following variables from SSP2015:

1. variable *q1_1*: gender
2. variable *age*: age
3. variable *q4*: educational level
4. variable *q6_1*: current employment status
5. variable *q31_1*: respondent's income
6. variabel *q31_2*: household income

It turns out that only the first four variables appears among the 114 variables that optimize LD, while *q31_1* and *q31_2* are excluded. The request of keeping *q31_1* and *q31_2* will lead to a sub-optimal solution necessarily, i.e. less data entries will be available for the final analysis. Therefore, we associate weights $\omega_i = 10$ to the above six variables, and weights $\omega_j = 1$ to all remaining variables. Moreover, we give equal weights to all respondents $W = \mathbb{I}_N$. The weighted optimization method leads to a solution of Eqs. (3) and (6) that selects 116 variables including also all the above six variables. However, LD deletion over all 116 variables leaves only 2637 respondents this time, corresponding to 50% of the dataset, corresponding to a 3.7% loss with respect to the optimization without weights. It is interesting to compare this result with the case where only those six variables are considered while all other variables are removed from the dataset. In this case, LD leaves 3182 respondents over six variables, i.e. a loss of 393 respondents only. Whether this case is better or worse than the one from the weighted optimization, depends mostly on the ultimate goal of the statistical analysis. There are situations where it may be advantageous to restrict the analysis to small subsets of the variables. In such cases, one must be careful when comparing results between different groups because LD may lead to *different* groups of respondents for different group of variables, which is an approach that has been strongly deprecated in the literature. In other situations, it may be preferable to work with the largest possible subset of the dataset, which is complete and can be used in the final statistical analysis. For instance, six variables and 3182 corresponds to 3.1% of the whole dataset, which is rather small when compared to the subset of 116 variables with 2637 respondents (about 50% of the whole dataset). Nonetheless, the weighted optimization of the LD method is sufficiently flexible to adapt to several situations reliably.

4 Conclusions

In this work we described how one can optimize the use of LD by maximizing the number of variables and respondents that remains after applying LD. The method is deterministic, and it provides a quantitative numerical guideline for selecting the optimal subset of variables. Moreover, the method allows also to prioritize groups of variables (and/or respondents) by means of weights in the optimization equations. We also perfected a general thresholding method that greatly helps the numerical implementation of the optimization algorithm. We tested it on several cases, and verified it is sufficiently powerful and robust to provide reliable results. Obviously, such a method should complement other heuristic approaches, and general considerations about the dataset. Although we do not have sufficient space to discuss all details here, we remind that one should always try to identify the reasons for missingness, in order to give a correct interpretation of the distribution of missing data, and to select the best method to analyze the dataset. Moreover, LD can be applied effectively without bias when data are missing completely at random, and it cannot substitute other methods (such as imputation methods) that are more effective when data are missing at random, or not at random (see, e.g. [1, 11]). Furthermore, our algorithm does not take into account the quality of the selected data, and therefore it could lead to inconvenient datasets. Nevertheless, our approach is sufficiently flexible to allow the inclusion of information on data quality, via the assignment of specific weight factors at the beginning of the optimization procedure.

Acknowledgements This work was supported by JSPS KAKENHI Grant Number 26380658, 17K04103, and 16H02045, as part of the SSP Project (http://ssp.hus.osaka-u.ac.jp). The authors thank the SSP Project for the permission to use the SSP 2015 survey. Finally, the authors would like to thank the anonymous Referees for their valuable and constructive comments.

References

1. Allison, P.D.: Multiple imputation for missing data: a cautionary tale. Sociol. Methods & Res. **28**(3), 301–309 (2000)
2. Allison, P.D.: Missing Data. Sage, Thousand Oaks (2001)
3. Berkovitz, L.: Convexity and Optimization in $\mathbb{R}^n$. Wiley, Hoboken (2002)
4. Cela, E.: The Quadratic Assignment Problem: Theory and Algorithms. Kluwer Academic, Dordrecht (1998)
5. Cook, D., Swayne, D.F.: Interactive and Dynamic Graphics for Data Analysis. Springer, New York (2007)
6. Ge, R., Huang, C.: A continuous approach to nonlinear integer programming. Appl. Math. Comput. **34**, 39–60 (1989)
7. King, G., Honaker, J., Joseph, A., Scheve, K.: Analyzing incomplete political science data: an alternative algorithm for multiple imputation. Am. Polit. Sci. Rev. **95**, 49–69 (2001)
8. Little, R.J.A., Rubin, D.B.: Statistical Analysis with Missing Data, 2nd ed. Wiley, New York (2002)

9. Murray, W., Ng, K.M.: An algorithm for nonlinear optimization problems with binary variables. Comput. Optim. Appl. **47**(2), 257–288 (2010)
10. National Research Council: The Prevention and Treatment of Missing Data in Clinical Trials. The National Academies Press, Washington (2010)
11. Schafer, J.L.: Analysis of Incomplete Multivariate Data. Chapman & Hill, London (1997)
12. Schafer, J.L., Graham, J.W.: Missing data: our view of the state of the art. Psychol. Methods **7**, 147–177 (2002)
13. SSP2015: 2015 Japanese Survey on Stratification and Social Psychology. http://ssp.hus.osaka-u.ac.jp
14. Wilkinson, L.: Statistical methods in psychology journals: guidelines and explanations. Am. Psychol. **54**(8), 594–604 (1999)

Part IV
Graphical Models

Measurement Error Correction by Nonparametric Bayesian Networks: Application and Evaluation

Daniela Marella, Paola Vicard, Vincenzina Vitale, and Dan Ababei

Abstract In this paper a procedure for measurement error correction based on nonparametric Bayesian networks is proposed. The performance of the proposed method is evaluated using a validation sample collected by Banca d'Italia and a major Italian bank group to investigate the measurement error mechanism in the main financial variables amounts observed in the Banca d'Italia survey on Household Income and Wealth. Specifically, in this paper attention is focused on the bond amounts. By means of Uninet's programmatic engine working directly from R, data can be corrected unit by unit by sampling from the nonparametric Bayesian network. Thanks to the validation sample, the distances between the true and the imputed values are computed and the procedure is evaluated.

Keywords Bayesian network · Imputation · Normal copula · Sampling

D. Marella
Dipartimento di Scienze della Formazione, Roma Tre University, Rome, Italy
e-mail: daniela.marella@uniroma3.it

P. Vicard (✉)
Dipartimento di Economia, Roma Tre University, Rome, Italy
e-mail: paola.vicard@uniroma3.it

V. Vitale
Dipartimento di Scienze Sociali ed Economiche, Sapienza University of Rome, Rome, Italy
e-mail: vincenzina.vitale@uniroma1.it

D. Ababei
LightTwist Software, Delft, Netherlands
e-mail: dan@lighttwist.net

F. Greselin et al. (eds.), *Statistical Learning of Complex Data*,
Studies in Classification, Data Analysis, and Knowledge Organization,
https://doi.org/10.1007/978-3-030-21140-0_16

1 Introduction

Variables observations are often affected by measurement error so that values observed in the data collection stage are different from the true ones. Measurement errors may lead to large bias effects on the estimation process. As a consequence, preliminary measurement error detection and correction should be performed before applying standard statistical inference techniques in order to avoid a serious impact on the survey results quality. To this aim, the error generating mechanism is modeled and estimated; then, microdata imputation is carried out. In this paper we focus attention on the respondent measurement error and we analyze and correct it using Bayesian networks.

When the variables are categorical, standard Bayesian networks (BNs, [2]) have been proposed as a tool to deal with measurement error. Simulations and applications are illustrated in [7] and [8]. For continuous variables a preliminary discussion is in [9].

In this paper we focus on continuous variables (such as, for example income, bond and share amounts) whose distributions is not necessarily Gaussian. The error generating mechanism is estimated using nonparametric Bayesian networks (NPBNs). In such a way, continuous data can be analyzed without any preliminary discretization process and unrealistic assumption of Gaussian distribution.

Moreover, an automatic procedure for measurement error correction based on sampling from the estimated NPBN is introduced and applied to a validation sample associated to the Banca d'Italia survey on Household Income and Wealth (SHIW, for short) and whose questionnaire and survey design are very close to those used in SHIW.

The paper is organized as follows. In Sect. 2 nonparametric Bayesian networks are briefly introduced. In Sect. 3 an application of nonparametric Bayesian networks to the validation sample provided to us by Banca d'Italia is illustrated. Results are shown in Sect. 3.1. Final discussion is in Sect. 4.

2 Nonparametric Bayesian Networks

BNs are multivariate statistical models satisfying sets of (conditional) independence statements contained in a directed acyclic graph (DAG). A DAG is a pair $G = (V, E)$ where V is the set of nodes and E is the set of directed edges between pairs of nodes. Each node represents a random variable, while missing arrows between nodes imply (conditional) independence between the corresponding variables. A directed graph is acyclic in the sense that it is forbidden to start from a node and, following arrows directions, go back to the starting node. Given that BNs are conditional independence models, they allow to describe and to read independencies from the DAG. There are properties connecting the concept of conditional independence between variables and absence of an arrow in the graph; these are encoded in the Markov properties. For more details, we refer to [6].

When the variables of interest are continuous, structural and parameter learning and evidence propagation can be performed under the assumption of Gaussian distribution [2]. However, in many real cases variables distributions are so far from normality that the Gaussian assumption becomes completely unrealistic. In such circumstances, continuous variables are generally discretized, and inference techniques for discrete BNs are used.

To avoid inappropriate assumption and discretization, multivariate nonlinear complex dependence structures can be modeled by copulas giving rise to NPBNs; for details, see [1] and [3].

Differently from standard BNs, in continuous NPBNs [5] nodes are associated with continuous invertible distribution functions and edges with (conditional) rank correlations that are realized by a chosen copula. Here the joint Gaussian copula is used. It satisfies the zero independence property: zero rank correlation is equivalent to zero partial correlation that, in turn, is equivalent to zero conditional correlation. Finally, the latter implies conditional independence and absence of the corresponding edge.

The main advantage of nonparametric networks is that the absence of an arc can be still interpreted as a (conditional) independence statement (as for standard BNs) so that the DAG can still be used to represent the conditional independence relations proper of a set of continuous variables of interest.

3 NPBNs and Measurement Error—Application and Results

In this paper, a NPBN based measurement error model for data on bond amounts in the survey on household income and wealth 2008 is estimated and a procedure for the detection and correction of measurement error is proposed. A validation sample, provided to us by Banca d'Italia, has been used to estimate the network and to evaluate the correction procedure performance.

SHIW is a biannual sample survey conducted by Banca d'Italia. Its main objective is to study the main sources of wealth (such as income, dwelling, investments in bonds, shares, and other financial products) of Italian households. Data in the validation sample have been collected through an independent experiment survey done by Banca d'Italia and a major Italian bank group on a sample of the latter. The survey was carried out in 2003 on a sample of 1681 households where at least one member was a customer of the bank group. Then survey data were matched with the bank customers database containing the amount of bonds and shares actually held by the statistical units selected in the sample.

A direct comparison of reported and true bond amounts shows that the 87% of household data are affected by underreporting; therefore, we proceed to correct bond amount declared values through a procedure based on NPBN. Since data quality suffers from a series of inconsistencies, preliminary data cleaning is carried out to improve data accuracy. In order to avoid inconsistencies ascribable to an individual having more than one bank account with investments in bonds, all

Table 1 Description of the variables analyzed in the NPBN model

Variable	Variable description
IV_AFNORISK	True amount of bonds
F_AFNORISK	Amount of bonds
F_AFRISK	Amount of stocks
ETA	Age of the head of the household
Y	Income
YM	Self-employment income
YLM	Payroll employment income
AR	Real assets
F_AF1	Certificates of deposit
F_AF2	Italian government securities (BOTs, CCTs, etc.)
F_AF3	Italian bonds and foreign securities
F_AF4	Quoted and not quoted shares
F_AF5	Mutual funds
F_AF6	Asset management
SUPAB	Surface of dwelling
LIE	Propensity to underreport

multi-banked customers (709) have been eliminated. Furthermore, 19 units with incoherent information are dropped out of the dataset. The final sample size is 844.

The list of the studied variables, together with their description, is reported in Table 1. All the variables, except the true bond amount owned by households ($IV\_AFNORISK$), are given by the values declared by the respondent. Differently from previous works [10], the variable *LIE* is continuous; it measures the propensity to underreport the true amount of bonds. We consider a respondent as a liar when the relative difference between the declared and the true bond amounts is greater than 10%, distinguishing misreporting in *bona fides*, attributable to an objective difficulty in retrieving the correct information, from the intentional one.

The analysis has been carried out using the software UniNet[1] where the joint Gaussian copula is used and the learning algorithm presented in [4] is implemented.

Notice that, since LIE is a continuous variable, the network structure can be learned directly from the overall set of data without the necessity to impose an arrow from $F\_AFNORISK$ to LIE, as done in [10] to overcome the problem of LIE being a binary variable.

In order to learn the structure, the variables are preliminarily ordered according to subject-matter knowledge and time/logical ordering. The NPBN is estimated starting from the saturated graph and computing all the rank and partial rank correlations associated to the edges. Next, those edges characterized by small rank correlations are removed.

[1] www.lighttwist.net/wp/uninet.

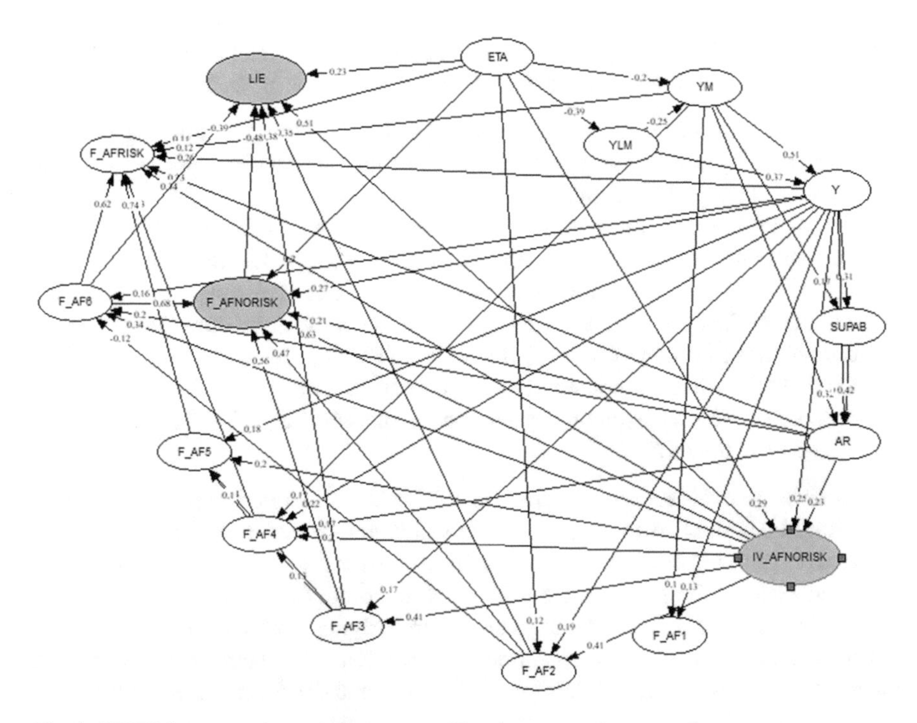

Fig. 1 NPBN for propensity to underreport and bond amount value correction

In our analysis of the validation sample, all edges with associated partial correlation larger than or equal to 0.1 are retained. The remaining rank correlations vary, in absolute value, from 0.10 to 0.74. The resulting NPBN is shown in Fig. 1. The network shows that the propensity to underreport, LIE, is directly influenced not only by the declared ($F\_AFNORISK$) and by the true ($IV\_AFNORISK$) bond amount, but also by age (ETA) and by some financial activities (F_AF2, F_AF3, F_AF6). Differently, income (Y) affects the underreport propensity indirectly only, i.e., *via* $F\_AFNORISK$, $IV\_AFNORISK$, F_AF2, F_AF3 and F_AF6.

The estimated NPBN has been validated using statistical tests based on the rank correlation matrices determinant and implemented in Uninet. Notice that all the above determinants take values in [0, 1] and are equal to 1 if all variables are independent, and equal to 0 if there is linear dependence between the variables. More specifically, the validation phase consists in two steps. The first one compares the determinant (DNR) of the empirical normal rank correlation matrix with that of the empirical rank correlation matrix (DER) to validate the joint normal copula. The second one compares the determinant (DBN) of the rank correlation matrix of the proposed NPBN with the determinant DNR to test if the estimated network is an adequate model of the saturated graph.

We use the NPBN in Fig. 1 to detect and correct measurement errors in the bond amounts. To this aim we propose the following three steps procedure:

1. Estimation of the propensity to underreport λ by inserting and propagating the evidence E given by the observed values of all variables except the true bond amount $IV\_AFNORISK$.
2. Estimation of the probability distribution of $IV\_AFNORISK$ given all the observed values by inserting and propagating the updated evidence, i.e., $(E, LIE = \hat{\lambda})$, throughout the network. The individual true value for $IV\_AFNORISK$ can then be predicted by a random draw from such a distribution and is denoted by $I\_AFNORISK$.
3. Computation of the imputed value. It coincides with the original one if $\hat{\lambda} < 0.2$, otherwise it is given by the linear combination: $\hat{\lambda}\, I\_AFNORISK + (1 - \hat{\lambda})\, F\_AFNORISK$.

In order to evaluate the performance of the above correction procedure, steps 1–3 have been applied to all units in the validation sample by means of Uninet's programmatic engine directly working from R (using the RDCOMClient library to connect to the engine). After setting the conditioning nodes to the observed values in the NPBN, Uninet calculates the underlying Gaussian conditional joint distribution analytically. The conditional distributions of the output nodes are then obtained by transforming the corresponding marginal distributions from Gaussians to their original ones. Notice that without this engine the conditionalization could be performed only by inserting a unit at time, making the overall dataset correction nearly impossible.

3.1 Results

Results arising from our imputation procedure, displayed in Table 2, are very promising. For comparison purposes the mean, in absolute value, of the distances between the true amount of bonds and the declared and the imputed ones respectively, are computed. As shown in Table 2, the proposed approach reduces the distance from the true values of 8.5%, on average.

From Table 2 it is also evident that the estimated network in Fig. 1 performs particularly well when it takes into account the liars only; for this group, the imputation procedure reduces the distance from the true values of 13.6%.

Table 2 Imputation procedure performance

Group	Distance true – observed value (Mean)	Distance true – imputed value (Mean)	Relative difference (%)
ALL	59594.23	54507.86	−8.5
LIARS	104456.8	90267.56	−13.6

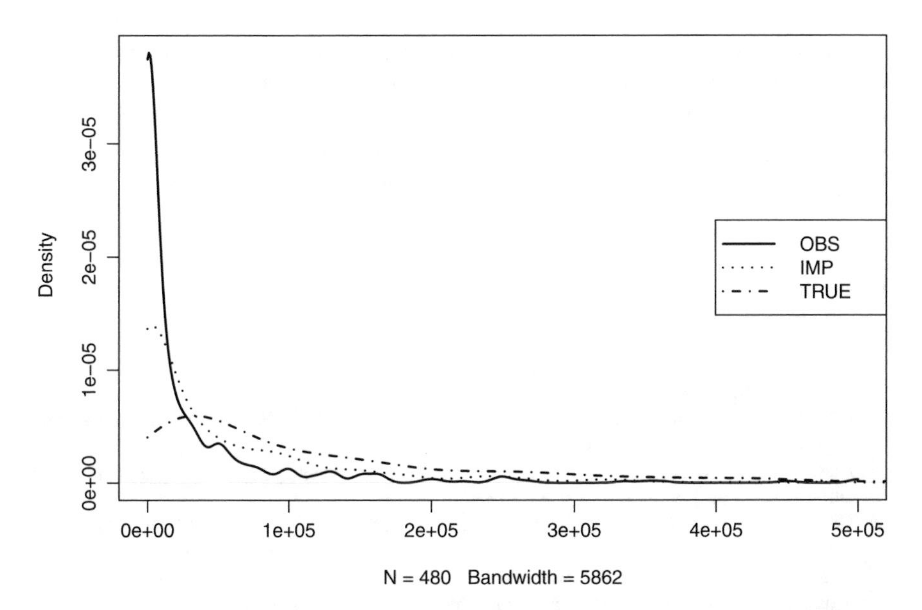

Fig. 2 Kernel density estimation of the observed, imputed, and true distribution of bonds—the group of liars

Finally, Fig. 2 shows the kernel density estimates of the observed (OBS), the imputed (IMP), and true (TRUE) amount of bonds respectively, for the liars only. The density of imputed values tends to reduce the number of low declared values with a distribution nearest to that of the true one. Analogous results are obtained when the whole sample is considered.

4 Conclusions

In this paper nonparametric Bayesian networks have been used to model the underreporting generating process affecting the Banca d'Italia Survey on household income and wealth. Moreover, a new procedure to sample from the NPBN and automatically correct the data has been introduced and evaluated on a validation sample associated to SHIW. The results are promising; therefore, the proposed imputation procedure could be a valid tool to deal with measurement errors when continuous variables with non-normal distribution, such as financial assets, are considered. A possible limitation of this analysis is given by the Gaussian copula assumption. Therefore further research could focus on the possible use of different copula families and on the consequent possibility to efficiently sample from the associated Bayesian network.

Another aspect deserving attention is that of the distribution shape of the variable to be corrected. As shown in Fig. 2, in our application the distribution is strongly asymmetric. As a consequence, the random draw from the conditional distribution may generate particularly large bond amounts, thus reducing the improvement due to imputation. In order to avoid this problem or to limit its effects, one solution could consist in identifying homogeneous groups with respect to their response behavior, and treat them separately in the imputation phase.

References

1. Bedford, T., Cooke, R.M.: Vines - a new graphical model for dependent random variables. Ann. Stat. **30**, 1031–1068 (2002)
2. Cowell, R.G., Dawid, A.P., Lauritzen, S.L., Spiegelhalter, D.J.: Probabilistic Networks and Expert Systems. Springer, New York (2007)
3. Hanea, A., Kurowicka, D., Cooke, R.: Hybrid method for quantifying and analyzing Bayesian belief nets. Qual. Reliab. Eng. Int. **22**, 613–729 (2006)
4. Hanea, A., Kurowicka, D., Cooke, R.M., Ababei, D.A.: Mining and visualising ordinal data with non-parametric continuous BBNs. Comput. Stat. Data Anal. **54**, 668–687 (2010)
5. Kurowicka, D., Cooke R.M.: Uncertainty Analysis with High Dimensional Dependence Modelling Methods for Bayesian Networks. Wiley, Chichester (2006)
6. Lauritzen, S.L.: Graphical Models. Oxford University Press, Oxford (1996)
7. Marella, D., Vicard, P.: Object-oriented Bayesian networks for modeling the respondent measurement error. Commun. Stat. Theory Methods **42**, 3463–3477 (2013)
8. Marella, D., Vicard, P.: Object-oriented Bayesian network to deal with measurement error in household surveys. In: Morlini, I., Minerva, T., Vichi, M. (eds.) Advances in Statistical Models for Data Analysis, pp. 157–164. Springer, Heidelberg (2015)
9. Marella, D., Vicard, P.: Towards an integrated Bayesian network approach to measurement error detection and correction. Commun. Stat. Simul. Comput. (2017). https://doi.org/10.1080/03610918.2017.1387664
10. Marella, D., Vicard, P., Vitale, V.: Non parametric Bayesian networks for measuremet error detection. In: Greselin, F., Mola, F., Zenga, M. (eds.) Cladag 2017. 11th Scientific Meeting of the Classification and Data Analysis Group. Book of short papers. Universitas Studiorum, Mantova (2017)

Copula Grow-Shrink Algorithm for Structural Learning

Flaminia Musella, Paola Vicard, and Vincenzina Vitale

Abstract The PC algorithm is the most known constraint-based algorithm for learning a directed acyclic graph using conditional independence tests. For Gaussian distributions the tests are based on Pearson correlation coefficients. PC algorithm for data drawn from a Gaussian copula model, Rank PC, has been recently introduced and is based on the Spearman correlation. Here, we present a modified version of the Grow-Shrink algorithm, named Copula Grow-Shrink; it is based on the recovery of the Markov blanket and on the Spearman correlation. By simulations it is shown that the Copula Grow-Shrink algorithm performs better than the PC and the Rank PC algorithms, according to the structural Hamming distance. Finally, the new algorithm is applied to Italian energy market data.

Keywords Gaussian copula · Rank-based correlation · Grow-Shrink algorithm

1 Introduction

A Bayesian network (BN, [2]) is a graphical model representing the multivariate probability distribution of a set of variables by means of a directed acyclic graph (DAG). BNs are applied in very many real contexts for their easy-to-read pictorial representation of complex problems and for their capability to evaluate scenarios. In fact, BNs can be provided with an inference engine allowing to carry out what-if

F. Musella
Department of Research, Link Campus University, Rome, Italy
e-mail: f.musella@unilink.it

P. Vicard
Dipartimento di Economia, Roma Tre University, Rome, Italy
e-mail: paola.vicard@uniroma3.it

V. Vitale (✉)
Dipartimento di Scienze Sociali ed Economiche, Sapienza University of Rome, Rome, Italy
e-mail: vincenzina.vitale@uniroma1.it

F. Greselin et al. (eds.), *Statistical Learning of Complex Data*,
Studies in Classification, Data Analysis, and Knowledge Organization,
https://doi.org/10.1007/978-3-030-21140-0_17

analysis by means of computationally efficient algorithms. However, building a BN can be tricky: most of times, the dependencies are unknown or partially known, so that the DAG can not be built manually but has to be estimated directly from data. Many structural learning methods are suitable for discrete data and for normal data, but only a few for non-Gaussian ones. For recent interesting developments about the use of DAG for non-normal data via copula function, we can refer to [1, 3]. Here, a Gaussian copula model is considered, thus a structural learning algorithm for nonparanormal data is proposed.

The paper is organized as follows: nonparanormal graphical models are briefly recalled in Sect. 2; known structural learning methods together with the proposed one are discussed in Sect. 3; simulation results are addressed in Sect. 4.1 while a real case application is shown in Sect. 4.2. Section 5 addresses the conclusions.

2 Nonparanormal Graphical Models

A DAG is a mathematical object made of a finite set of nodes and directed edges arranged without producing directed cycles. Nodes of a DAG are associated with random variables, either discrete or continuous, and arrows between nodes represent direct relevance of one variable to another. A node, say X_i, is said parent of another node, say X_j, if there is an arrow from X_i to X_j; correspondingly, X_j is said child of X_i. The DAG is thus a map of conditional independence statements that can be read by means of the d-separation criterion [13]. DAGs entailing the same set of conditional independence relations are called Markov equivalent and can be represented by a Partially DAG (PDAG) or, uniquely, by a Completed Partially DAG (CPDAG).

Recently, nonparanormal graphical models have been defined in [8] as follows:

Definition 1 Let $f = (f_v)_{v \in V}$ be a collection of strictly increasing functions $f_v : R \rightarrow R$ and $\Sigma \in R^{V \times V}$ be a positive definite correlation matrix. The nonparanormal distribution $NPN(f, \Sigma)$ is the distribution of the random vector $(f_v(Z_v))_{v \in V}$ for $(Z_v)_{v \in V} \sim N(0, \Sigma)$.

Definition 2 The nonparanormal graphical model $NPN(G)$ associated with a DAG G is the set of all distributions $NPN(f, \Sigma)$ that are Markov with respect to G.

The function f_v realizes a deterministic transformation on Z_v preserving the same dependence structure of the underlying latent multivariate normal distribution in the nonparanormal model.

If $X \sim NPN(f, \Sigma)$ and $Z \sim N(0, \Sigma)$, then $X_A \perp\!\!\!\perp X_B | X_S \Leftrightarrow Z_A \perp\!\!\!\perp Z_B | Z_S$, for any triple of pairwise disjoint sets $A, B, S \subset V$. For two nodes (u, v) and a separating set S we have $X_u \perp\!\!\!\perp X_v | X_S \Leftrightarrow \rho_{uv|S} = 0$. Accurate estimators for latent normal correlation coefficients are produced by a trigonometric transformation on Spearman rank correlation (r). Reference [6] shows that if (X, Y) is a bivariate normal with $Corr\,(X, Y) = \rho$, then:

$$P\left(|2sin\left(\frac{\pi}{6}\widehat{r}\right) - \rho| > \varepsilon\right) \leq 2exp\left(-\frac{2}{9\pi^2}n\varepsilon^2\right) \tag{1}$$

Since $\widehat{r}$ depends on the observations *via* their ranks that are preserved under strictly increasing functions, (1) still holds for nonparanormal graphical models with Pearson correlation $\rho = \Sigma_{xy}$ in the underlying latent bivariate normal distribution. Therefore, ρ is estimated by Pearson formula [9] as:

$$\hat{\rho} = 2\sin\left(\frac{\pi}{6} \cdot \hat{r}\right) \tag{2}$$

The same transformation still holds for the partial correlation coefficients.

3 Structural Learning

The BN learning process consists of two phases: building the DAG corresponding to the conditional independence statements, and estimating marginal and conditional probability distributions. Most of times the networks have to be learned from data. In such situation computationally efficient algorithms are needed. Structural learning methods are mainly developed according to two approaches: *scoring and searching* techniques spanning the space of models and selecting those optimizing an information criterion, and *constraint-based* algorithms using conditional independence tests. The main constraint-based methods are briefly presented below.

PC Algorithm and Rank PC Algorithm
The PC algorithm [11] is a backward algorithm consisting of three steps:

1. identification of the skeleton of the graph (i.e., the underlying undirected graph) by recursively testing marginal and conditional independencies;
2. identification of *v*-structures—three nodes configurations such as $X_i \rightarrow X_k \leftarrow X_j$, standing for conditional dependence between X_i and X_j given X_k—on the basis of the test results of the previous step;
3. orientation of the remaining undirected links without producing additional *v*-structures and/or directed cycles.

For multivariate normal observations, in [4] it is shown that PC algorithm has high-dimensional consistency properties. In case of normal data, PC algorithm tests conditional independence between two variables, say X and Y given a separating set S, by computing the partial correlation $\rho_{X;Y|S}$. It holds that: $X \perp\!\!\!\perp Y|S \Leftrightarrow \rho_{X;Y|S} = 0$. The sample partial correlation $\hat{\rho}_{X;Y|S}$ is used as a good estimate of $\rho_{X;Y|S}$.

In many situations variables are not Gaussian, then a PC algorithm rank version, named Rank PC (RPC) algorithm, has been proposed by Naftali and Drton [8]. RPC algorithm tests conditional independence between two variables, say X and Y, given a separating set S by computing the rank-based partial correlation estimates between X and Y in (2). The RPC algorithm consistency is proved in [8] under some non-strict assumptions. It is shown that RPC works at the same strength of PC algorithm for normal data, but considerably better for non-normal data under the assumption of joint distribution following a normal copula model. The PC and RPC algorithms are implemented in `pcalg` `R` package [5].

Grow-Shrink Algorithm and Copula Grow-Shrink Algorithm
The Grow-Shrink algorithm (GS, [7]) uses the concept of Markov blanket of a variable. The Markov blanket of a node X, MB(X), consists of all parents, children, and parents of children of X. MB(X) d-separates X from any other variable outside MB(X). In other words, MB(X) contains all the variables in the graph carrying information about X. The GS algorithm focuses on the recovery of MB(X) using pairwise independence tests. It consists of two phases. In the first phase (growing), MB(X) is initially an empty set, denoted by S. Then the algorithm adds variables to S as long as they are associated with X given the current contents of S. In this phase, even variables not really belonging to MB(X) could be added to S. The second phase (shrinking) is performed to identify and remove these variables. The GS is implemented in `bnlearn` `R` package [10].

Here the Copula Grow-Shrink (CGS) algorithm is proposed. It has the same logical structure as GS but the marginal and partial correlations coefficients used in the statistical test for independence are computed through (2). CGS algorithm allows to learn the structure even when data are non-Gaussian. In this way the unrealistic normality assumption and preliminary data discretization can be avoided.

4 Experiments

In this section the CGS performance is discussed both in comparison with other algorithms by a simulation (Sect. 4.1) and by an application to real data (Sect. 4.2).

4.1 Simulation

According to the procedure implemented in the `pcalg` `R` package [5], and to ensure faithfulness (i.e., the exact correspondence between conditional independencies in the distribution and in the DAG), a random DAG, made of ten nodes and with sparsity parameter $s = 0.4$, is simulated (see Fig. 1). Data are sampled from it following: (a) multivariate normal distribution; (b) Gaussian Copula distribution, whose latent multivariate normal distribution is that of (a); (c) contaminated data from a mixture of Gaussian (80) and Cauchy (20) distributions not belonging to the nonparanormal models, as in [8]. We sample 250 distributions for each type

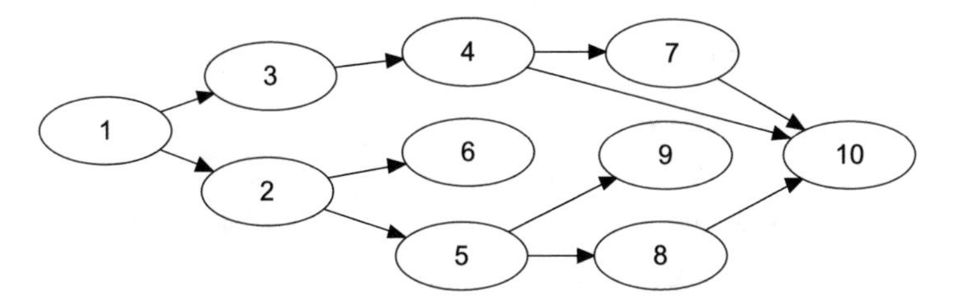

Fig. 1 The simulated graph

Table 1 Simulation results by sample size, data type and algorithm

		n = 50			n = 500		
Data type	Algorithm type	SHD (mean)	SHD (sd)	SHD (IQR)	SHD (mean)	SHD (sd)	SHD (IQR)
Normal data	PC	7.56	1.76	3	3.57	2.34	4
	RPC	7.81	1.69	2	3.69	2.34	4
	GS	5.30	1.51	2	1.89	1.07	1
	CGS	5.99	1.72	2	2.25	1.71	1
Non gaussian copula	PC	7.77	1.47	2	5.50	2.34	3
	RPC	7.67	1.95	3	4.01	3.05	7
	GS	6.92	1.82	2	4.22	2.20	3
	CGS	5.95	1.84	3	1.49	2.27	2
Gaussian copula	PC	8.65	1.72	3	5.87	2.02	2
	RPC	7.81	1.69	2	3.69	2.34	4
	GS	6.38	1.69	2	4.08	2.20	3
	CGS	5.99	1.72	2	2.25	1.71	1

of data described above, fixing $n \in \{50, 500\}$. On every training set, structural learning has been performed using PC, RPC, GS, and CGS algorithms with a 0.01 significance level. Algorithm performances have been compared in terms of *structural Hamming distance* (SHD, [12]), a measure counting the number of actions (add, delete, reverse) necessary to transform the estimated graph into the true one.

In Table 1 the SHD mean and standard deviation relative to all simulations are reported. For small sample size ($n = 50$), results show that the mean value is always smaller for the CGS algorithm than for PC, RPC, and GS when data are not normal; the standard deviation is instead quite unstable due to the presence of outliers (see Fig. 2a). Box-plots comparison shows that outliers, present in CGS box-plot, correspond to non-anomalous values in the PC and RPC distributions. For large sample size ($n = 500$), with the exception of normal data, the mean value is always much smaller for CGS than for PC, RPC, and GS algorithms. The CGS distribution is concentrated on SHD values smaller than those for PC, RPC, and GS distributions (box-plot in Fig. 2b). CGS box-plot shows many outliers, but they correspond to non-anomalous values in the other algorithm distributions (Fig. 2b).

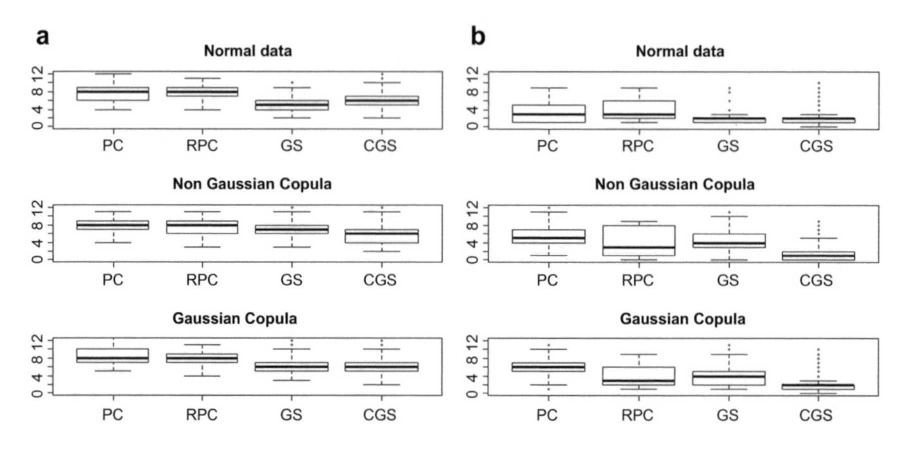

Fig. 2 Box-plot of SHD performance by algorithms and data typology for $n = 50$ (**a**) and $n = 500$ (**b**)

4.2 An Application to Real Data: The Italian Energy Market

Here, the CGS algorithm is applied to real data provided by the most important multi-national power company in Italy, the Enel group. The energy production cost is generally directly proportional to energy demand: as demand increases, the number of energy plants has to be upgraded accounting for production plants efficiency (first the most efficient, less expensive, then the less efficient, more expensive and sometimes polluting). Variability in the energy demand, due to seasonal fluctuations, causes uncertainty in identifying the optimal amount of production. Statistically speaking, the association structure among the variables of interest constitutes a relevant information for energy managers, and BNs can be a useful tool to estimate these relationships.

Data are referred to the Italian energy market and their monthly mean values span from January, 2014 to May, 2017.[1] As shown in Table 2, the variables involved in the analysis concern the energy price and its demand, the main important energy commodities (out of which Hydro, Wind, Solar, and Geothermal are renewable), and their costs.[2] The variables distributions are non-Gaussian, then a structural learning algorithm for non-normal data, like CGS, is needed.

The CGS algorithm allows to incorporate some prior knowledge in the structure learning process. Here, the following arc directions are forbidden: *Gas* and *Coal* production towards all renewable energies and towards the *Demand*. The resulting graph, for $\alpha = 0.05$, is shown in Fig. 3.

---

[1]As requested by Enel experts, we did not treat the time series before modelling them, since the aim is catching and modeling the variability of the energy market as a whole, including variables' seasonality and non stationarity.

[2]Hereafter, node names will coincide with variable names and will be written in italic.

Table 2 Description of variables involved in the analysis

Name	Description
Demand	Average total monthly energy demand (in MWh)
Hydro	Average monthly energy produced by hydroelectric power plants (in MWh)
Wind	Average monthly energy produced by wind power plants (in MWh)
Solar	Average monthly energy produced by solar power plants (in MWh)
Geothermal	Average monthly energy produced by geothermal power stations (in MWh)
Gas	Average monthly energy produced by thermal power plants burning gas (in MWh)
Coal	Average monthly energy produced by thermal power plants burning coal (in MWh)
Others	Average monthly energy produced by other power plants (in MWh)
Gas cost	Average monthly gas price (in Euro per MWh)
Coal cost	Average monthly coal price (in Euro per MWh)
Energy price	Average monthly energy price (in Euro per MWh)

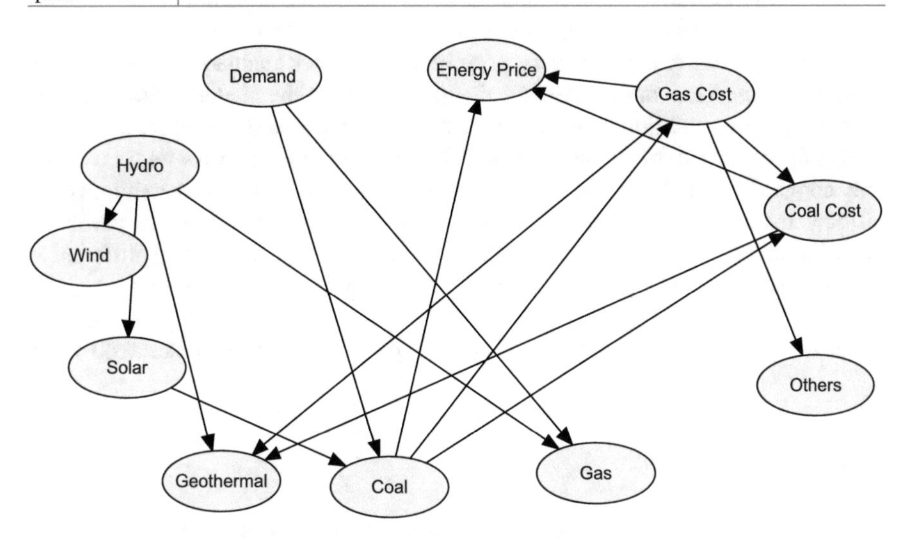

Fig. 3 DAG for the Italian energy market learned by CGS algorithm

The model seems to well reflect the dependence structure of the Italian energy market. It is known that, in Italy, the main power sources are natural gas and hydroelectricity; the national energy plan also includes an increasing power generation from all other renewable sources. In particular, Italy is among the largest producers of electricity from solar energy; wind and geothermal powers also give a contribution to satisfy the national energy demand. All these features are clearly depicted by the estimated network structure. In fact, the *Hydro* production influences all the other renewable sources and the *Gas* production levels. As expected, the energy *Demand* has a direct effect on *Coal* and *Gas* production but not at all on the renewable energy productions since their levels depend on weather and seasonality only. The *Energy Price* is directly affected by the *Coal* production and by the costs of both non-renewable energy commodities (*Coal Cost* and *Gas Cost*). All other variables influence the energy price indirectly; for instance, *Demand* impacts on *Energy Price via Coal* production.

5 Conclusions

In this paper the issue of BN structural learning for nonparanormal data has been addressed. The Copula Grow-Shrink algorithm is proposed, based on the recovery of the Markov blanket of the nodes and on the Spearman correlation. The paper provides both a simulation study to compare, in terms of learning performance, the proposed CGS algorithm to PC, RPC, and GS algorithms, and an application to the Italian energy market data. The application shows that the algorithm is appropriate for catching relations in a complex field such as that of an energy market.

The simulation results are very promising and other evaluations are going to be done in the next future. Among these: comparing the skeleton identification ability by using additional performance measures such as TPR, FPR, and TDR; analyzing the reason of the outliers in CGS distribution to improve its robustness; and comparing simulation results for different levels of DAG sparsity. As for all constraint-based algorithms, the sensitivity to the α value deserves particular attention since there might be a lack of robustness with respect to different α specifications. Therefore, further research will be devoted to this aspect.

Acknowledgements The research has been financed by Enel group (contract code: 8400067126).

References

1. Bauer, A., Czado, C.: Pair-copula Bayesian networks. J. Comput. Graph. Stat. **25**(4), 1248–1271 (2016)
2. Cowell, R.G., Dawid, P., Lauritzen, S.L., Spiegelhalter, D.J.: Probabilistic Networks and Expert Systems. Springer, New York (1999)
3. Elidan, G.: Copula Bayesian networks. In: Lafferty, J.D., Williams, C.K.I., Shawe-Taylor, J., Zemel, R.S., Culotta, A. (eds.) Advances in Neural Information Processing Systems, vol. 23, pp. 559–567. Curran Associates, Inc., Red Hook (2010)
4. Kalisch, M., Bühlmann, P.: Estimating high-dimensional directed acyclic graphs with the pc-algorithm. J. Mach. Learn. Res. **8**, 613–636 (2007)
5. Kalisch, M., Mächler, M., Colombo, D., Maathuis, M.H., Bühlmann, P.: Causal inference using graphical models with the R package pcalg. J. Stat. Softw. **47**(11), 1–26 (2012)
6. Liu, H., Han, F., Yuan, M., Lafferty, J., Wasserman, L.: High-dimensional semiparametric gaussian copula graphical models. Ann. Stat. **40**(4), 2293–2326 (2012)
7. Margaritis, D.: Learning Bayesian Network Model Structure from Data. Ph.D. thesis, School of Computer Science, Carnegie-Mellon University, Pittsburgh, PA (2003). Technical Report CMU-CS-03-153

8. Naftali, H., Drton, M.: Pc algorithm for nonparanormal graphical models. J. Mach. Learn. Res. **14**, 3365–3383 (2013)
9. Pearson, K.: Mathematical contributions to the theory of evolution - XVI. On further methods of determing correlation. Draper's Company Research Memoirs. Biometric Series, vol. 4, pp. 1–39 (1907)
10. Scutari, M.: Learning Bayesian networks with the bnlearn R, package. J. Stat. Softw. **35**(3), 1–22 (2010)
11. Spirtes, P., Glymour, C., Scheines, R.: Causation, Prediction, and Search, 2nd edn. MIT press, Cambridge (2000)
12. Tsamardinos, I., Brown, L.E., Aliferis, C.F.: The max-min hill-climbing Bayesian network structure learning algorithm. Mach. Learn. **65**(1), 31–78 (2006)
13. Verma, T., Pearl, J.: Equivalence and synthesis of causal models. In: Proceedings of the Sixth Annual Conference on Uncertainty in Artificial Intelligence, UAI '90, pp. 255–270. Elsevier Science Inc., New York (1991)

Context-Specific Independencies Embedded in Chain Graph Models of Type I

Federica Nicolussi and Manuela Cazzaro

Abstract For a set of variables collected in a contingency table, we focus on a particular kind of relationships such as the context-specific independencies. These are conditional independencies that hold for particular values of the conditioning set. Given the advantages of the graphical models, we use them to represent different relationships among the variables, including the context-specific independencies. In particular, we enrich chain graph models with labelled arcs. Furthermore, we consider the well-known relationships between chain graph models and hierarchical multinomial marginal models and we introduce new constraints on parameters in order to describe the context-specific relationship. Finally, we provide an application to the study of innovation in Italy by comparing two different periods.

Keywords Context-specific independencies · Categorical variables · Ordinal variables · Stratified chain graph models

1 Introduction

A context-specific independence (CSI) is a particular relationship that focuses on certain value(s) of conditioning variables. Indeed, it is not rare to observe phenomena that are independent under particular conditions, but, under other circumstances, they have on the contrary a strong connection. In this case, stating that there is conditional independence between the two phenomena is not true, but not considering the lack of "partial" connection could be inaccurate. In this work we

F. Nicolussi (✉)
Department of Economics, Management and Quantitative Methods, University of Milano, Milano, Italy
e-mail: federica.nicolussi@unimi.it

M. Cazzaro
University of Milano-Bicocca, Milano, Italy
e-mail: manuela.cazzaro@unimib.it

F. Greselin et al. (eds.), *Statistical Learning of Complex Data*,
Studies in Classification, Data Analysis, and Knowledge Organization,
https://doi.org/10.1007/978-3-030-21140-0_18

consider a set of categorical variables and we study different kind of relationships, among which the CSIs that lie between them. In the literature, marginal and conditional independencies get more attention and are deeply studied. For instance, given two variables, say X_1 and X_2, it is usual to investigate if they are marginally independent ($X_1 \perp X_2$) or conditionally independent given a third variable X_3 ($X_1 \perp X_2|X_3$). The CSI statement establishes that the variables X_1 and X_2 are independent given $X_3 = i_3$, while the same statement does not hold when $X_3 \neq i_3$; see among others Boutilier [2]. Indeed, these CSIs were mainly examined to study problems concerning latent variables; see, for instance, [13].

Our aim is to incorporate the CSI conditions in graphical models that are suitable to represent different kind of relationships. Nyman et al. [11, 12] analyse CSIs in graphical models based on undirected graphs, or on directed acyclic graphs, using the classical log-linear parametrization. In both papers they adapt these kind of graphs with labelled arcs in order to take into account the CSIs. In this work we follow the same approach and we enrich chain graphs with labelled arcs in order to display also the CSIs. Furthermore, we take advantage of hierarchical multinomial marginal (HMM) parametrization [1, 4], as a generalization of the log-linear models, to represent the dependence relationships. A further advantage to consider the CSIs lies also in the possibility of reducing the number of HMM parameters.

This paper has the following structure: In Sect. 2.1 we give an overview of HMM models by considering also the case when we deal with ordinal variables. About this, we propose the constraints on HMM models able to satisfy the CSIs. The representation through (stratified) chain graph models is debated in Sect. 2.2. In Sect. 3 an application on the study of the trend of innovation degree on Italian enterprises is provided. Finally, Sect. 4 is dedicated to a conclusion.

2 Methodology

Let us consider q categorical (ordinal) variables $V = \{X_1, \ldots, X_q\}$ taking values in the contingency table $\mathcal{I} = (\mathcal{I}_1 \times \cdots \times \mathcal{I}_q)$, where $\mathcal{I}_j = \{1, \ldots, n_j\}$, with $j = 1, \ldots, q$, such that $i_j \in \mathcal{I}_j$ is the generic value of the variable X_j. Note that $(i_1, \ldots, i_q)$ identifies a particular cell of the contingency table $\mathcal{I}$ and henceforth we refer to it with the shortcut $(i_{1\ldots q})$. In the following subsection we describe the methodology able to define a system of independencies (marginal, conditional and context specific) that reveals the relationships among all the variables involved in the contingency table.

2.1 *Hierarchical Multinomial Marginal Models for Context-Specific Independencies*

The HMM model is a generalization of the classical log-linear model which allows to represent conditional and marginal independencies in the same model. Instead of considering only the joint distribution, this model takes into account also marginal distributions and, on these, defines the log-linear parameters by respecting certain properties of completeness and hierarchy. These new parameters are contrasts (of sum) of logarithms of probabilities and henceforth we refer to them as HMM parameters.

Let us consider 3 variables, X_1, X_2 and X_3. We are interested in describing that variables X_1 and X_2 are independent given by X_3, jointly considered, $X_1 \perp X_2|X_3$, and that X_2 is marginally independent of X_3, $X_2 \perp X_3$. To this aim, we consider the marginal distribution of $\{X_2, X_3\}$ and the joint distribution. We refer to these by defining the class of marginal distributions $\{\{2, 3\}; \{1, 2, 3\}\}$, where $\{2, 3\}$ and $\{1, 2, 3\}$ are a shortcut for $\{X_2, X_3\}$ and $\{X_1, X_2, X_3\}$. Then we define the classical log-linear parameters on the marginal contingency table $\mathscr{I}_{23}$ concerning the variables $\{2, 3\}$ and the remaining parameters on the contingency table $\mathscr{I}$. Let us define the HMM parameters with the caption $\eta^{\mathscr{M}}_{\mathscr{L}}(i_{\mathscr{L}})$, where $\mathscr{M}$ refers to the marginal distribution, $\mathscr{L}$ denotes the subset of variables to which the parameter pertains and $i_{\mathscr{L}}$, in parentheses, represents the values of the variable selected in $\mathscr{L}$ (when the parentheses are omitted, it means that the parameters refer to each $i_{\mathscr{L}} \in \mathscr{I}_{\mathscr{L}}$). Finally, in order to test the marginal and conditional independencies, we have to constrain to zero the parameters $\eta^{\{2,3\}}_{2,3}$, $\eta^{\{1,2,3\}}_{1,2}$ and $\eta^{\{1,2,3\}}_{1,2,3}$.

Let us consider the following statement of CSI where the conditional independence holds only in a subset of variables. For instance,

$$\begin{cases} X_1 \perp X_2|X_3 = i_3,\ i_3 \in \mathscr{K} \\ X_1 \not\perp X_2|X_3 = i_3,\ i_3 \notin \mathscr{K}, \end{cases} \tag{1}$$

where $\mathscr{K} \subseteq \mathscr{I}_3$ is a subset of the values i_3 of X_3 for which the conditional independence holds.

One main goal of this work is the definition of the constraints on HMM parameters in order to satisfy the CSI in formula (1). In [11], Nyman et al. deal with the log-linear parameters defined on the joint distribution. Here, as first improvement, we take into account the HMM parameters defined also on marginal distributions, see [10]. Thus, as before, we proceed to define the class of marginal distributions (the same mentioned above) and to specify the parameters evaluated on suitable marginal distributions. The constraints satisfying the CSI in formula (1) are

$$\eta^{\{1,2,3\}}_{1,2}(i_{12}) + \eta^{\{1,2,3\}}_{1,2,3}(i_{12}, i_3) = 0 \qquad i_{12} \in \mathscr{I}_{12} \quad i_3 \in \mathscr{K}, \tag{2}$$

where $\mathscr{I}_{12}$ is the marginal contingency table concerning the variables $\{1, 2\}$. Another important aspect of this work is to consider the possible presence of ordinal variables. The classical log-linear models, in fact, look poor when we want to focus on the interpretation of the effects among the variables, in particular, when we take into account ordinal variables; see, for instance, [3]. For this reason we choose different criteria for coding the variables through the parameters. In fact, beyond the classical *baseline* criterion, we take advantage of the *local* criterion that is more suitable for ordinal variables. By adopting the *local* criterion for coding the conditioning variable, as it is shown in [10], the constraints in formula (2) become

$$\eta_{1,2}^{\{1,2,3\}}(i_{12}) + \sum_{i_3^*=1}^{i_3} \eta_{1,2,3}^{\{1,2,3\}}(i_{12}, i_3^*) = 0 \qquad i_1 \in \mathscr{I}_{12} \quad i_3 \in \mathscr{K}. \tag{3}$$

It is worthwhile to note that, when we deal with local parameters, if the CSI is presented in the following different statement: $X_1 \perp X_2|X_3 \geq i_3$, $i_3 \in \mathscr{K}$, the constraints in formula (3) are equal to the ones in formula (2). More details are given in [10].

2.2 *Stratified Chain Graph Models*

A *chain graph* (CG) is a graph with both directed and undirected arcs and without any directed or semi-directed cycle. The vertices of a CG can be grouped in the so-called *chain components*, denoted by $T_1, \ldots., T_s$, that are the connected undirected components. Intuitively, chain graph models (CGMs) are graphical models which take advantage of chain graphs; see [5]. The structure of relationships among variables which follow an inherent order is well represented from these models. In particular, we can distinguish variables linked by symmetric relationships and variables linked by unilateral dependence. In this case we follow this order for collecting them in chain components.

In the literature, the representation of independencies through CGs is not unique, a deep dissertation is discussed in [5]; in this work, we adopt the point of view of Lauritzen and Wermuth [7], also known as chain graph models of type I, that is a subclass of the HMM models; see [9] and [14]. These CGMs are the natural extension of the graphical models based on undirected graph and directed acyclic graph. They interpret the lack of (un)directed arcs conditionally with respect to the remaining variables in the same component. In addition, all the systems of independencies representable through these graphical models benefit from the existence of a smooth likelihood function. In order to take into account the CSIs, we propose stratified chain graph models (SCGMs) as extension of stratified graphical models (SGMs) introduced by Nyman et al. [11]. Similarly to SGM, we denote the CSIs through labelled arcs. Figure 1 depicts an example of a SCGM. In this case the lack of the directed arcs between the nodes X_1 and X_5, X_2 and X_5 and finally between X_2 and X_3 represents the conditional independencies $X_1X_2 \perp X_5|X_3X_4$ and $X_2 \perp X_3|X_1X_4X_5$. Then the labelled arc between the nodes X_3 and X_4 represents the CSI $X_3 \perp X_4|X_1X_2X_5 = (i_1, *, i_5)$, where the asterisk is a symbol for referring to all the values of the variable X_2 in this case.

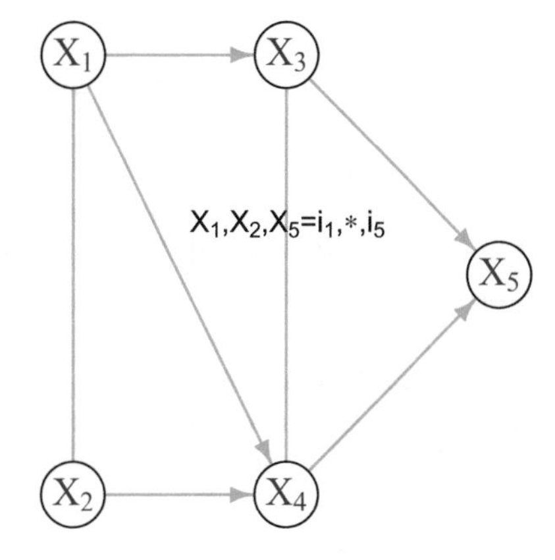

Fig. 1 SCGM with the labelled arc $X_3 - X_4$ referring to value i_1 of X_1, value i_5 of X_5 and all values of variable X_2

3 Application

In the next subsection we implement the presented model with an application to a real dataset. At first, we select the variables and we define the marginal distributions to take into account, according to the focus of the analysis. In order to find the best fitting model, we proceed with a three-step algorithm where each model is tested by using the likelihood ratio test G^2. The algorithm is explained below.

Step 1 We test the CGMs associated to all possible CGs obtained by deleting only one arc (at time) from the complete graph. Among these models, we select the ones with a *p-value* of the likelihood ratio test greater than 0.01.

Step 2 Similarly to Step 1, we test the SCGMs associated to all possible SCGs obtained by replacing only one arc (at time) with a labelled arc with all possible labels considered one at time. Among these models, we select the ones with a *p-value* of the likelihood ratio test greater than 0.1.

Step 3 From all admissible models selected in the previous two steps, we test all possible combinations of marginal, conditional independencies and CSIs and we maintain the one with lower AIC (Akaike information criterion) between the models with a *p-value* higher than 0.05.

3.1 The Italian Innovation Survey

We analyse two datasets, concerning the Italian Innovation Survey, pertinent each to a 3-year period: the first 2008–2010 and the second 2010–2012 [6]. The two datasets involve 16,531 and 18,697 small and medium sized Italian firms, respectively. We evaluate the revenue growth between the considered years, X_1 (1 = No, 2 = Yes). Then, we consider different factors that contribute to the innovation status of an enterprise: *innovation in products or services or production line or investment in*

R&D, X_2 (1 = No, 2 = Yes); *innovation in organization system*, X_3 (1 = No, 2 = Yes) and *innovation in marketing strategies*, X_4 (1 = No, 2 = Yes). Another type of variables we consider concerns the firm's features: the *main market (in revenue terms)*, X_5 (A = Regional, B = National, C = International); the *percentage of graduate employers*, X_6 ($1 = 0\% \vdash 10\%$, $2 = 10\% \vdash 50\%$, $3 = 50\% \vdash 100\%$) and the *enterprise size*, X_7 (1 = Small, 2 = Medium). We consider three marginal distributions. First, let us define the marginal distribution $\{5, 6, 7\}$ in order to study the symmetric relationships among the firm features; the second distribution $\{2, 3, 4, 5, 6, 7\}$ to highlight possible influences of the firm features on the innovation variables; finally, we consider the joint distribution $\{1, 2, 3, 4, 5, 6, 7\}$ in order to point out the effect of all variables on the revenue growth.

Following the three-step algorithm proposed in Sect. 3, in Figs. 2 and 3, we report the best fitting SCGM for the period 2008–2010 and 2010–2012, respectively. Note that the CSIs are represented by red arcs.

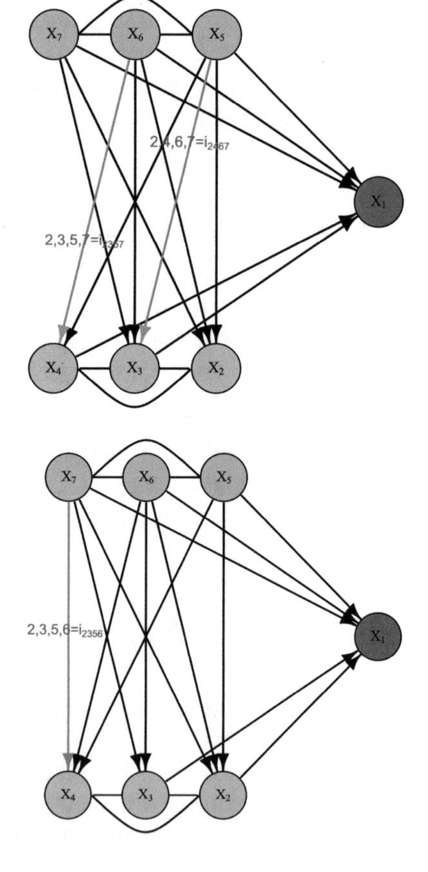

Fig. 2 Best fitting SCGM for the period from 2008 to 2010. $i_{2357} = (2, 2, 3, *)$ and $i_{2467} = (1, 2, 3, 2)$

Fig. 3 Best fitting SCGM for the period from 2010 to 2012. $i_{2356} = (2, 2, *, 3)$

Table 1 Values of the statistic tests of the selected HMMMs corresponding to the SCGMs in Fig. 2 (period 2008–2010) and Fig. 3 (period 2010–2012) with the list of independencies that they represent

Period	Independencies	G^2	df	p-Value	AIC
2008–2010	$X_1 \perp X_2 \| X_3X_4X_5X_6X_7$	126.02	112	0.17	−225.98
	$X_4 \perp X_7 \| X_2X_3X_5X_6$				
	$X_4 \perp X_6 \| X_2X_3X_5X_7 = i_{2357}$				
	$X_3 \perp X_5 \| X_2X_4X_6X_7 = i_{2467}$				
2010–2012	$X_1 \perp X_4 \| X_2X_3X_5X_6X_7$	145.93	123	0.08	−184.06
	$X_3 \perp X_5 \| X_2X_4X_6X_7$				
	$X_4 \perp X_7 \| X_2X_3X_5X_6 = i_{2356}$				

where $i_{2357} = (2, 2, 3, *)$, $i_{2467} = (1, 2, 3, 2)$ and $i_{2356} = (2, 2, *, 3)$

The list of independencies underlying the two SCGMs, together with the likelihood ratio test G^2, the corresponding *p-value* and the AIC value, is reported in Table 1.

Note that, in the two figures, despite the same structure of the undirected arcs, the presence (absence) of the directed arcs changes a little bit. In particular, in terms of innovation, the variable X_3 affects the growth X_1 in both models, while the influence of the other two innovation variables, X_2 and X_4, interchanges. Furthermore, the dependence relationships between the variables X_7 and X_4 or between X_5 and X_3 result weak or null. Indeed, these independencies are present in both models, under the conditional or the CSI point of view. In the first period we may found the additional CSI between the X_6 and the X_4.

Focusing on the CSIs, we recognize in the first model that the percentage of graduate employers (X_6) does not affect the innovation in marketing strategies (X_4) when there is an innovation in products or services ($X_2 = 2$) and in the organization system ($X_3 = 2$) and when, whatever the size of the company ($X_7 = *$), the firm works mainly in an international market ($X_5 = 3$). Again, we can recognize that the type of the main market where the firm operates (X_5) does not affect the innovation in the organization system (X_3) when there is no innovation in products and services ($X_2 = 1$) but there is innovation in marketing strategies ($X_4 = 2$), the percentage of graduate employers is high ($X_6 = 3$) and the enterprise size is medium ($X_7 = 2$). On the other hand, in Fig. 3 we can see that the size of the firm (X_7) does not affect the innovation in marketing strategies (X_4) when there is innovation in both products and services ($X_2 = 2$) and organization system ($X_3 = 2$), for any kind of market ($X_5 = *$) and when the employers are highly specialized (high degree of graduated employers, $X_6 = 3$).

All the analyses are carried out with the statistical software R, and the `package` `hmmm` [4].

4 Conclusions

The representation of relationships among categorical variables increases by considering the CSIs. Graphical models have shown useful properties in the representation of complex structure of dependencies and, also in this case, they reveal suitable features. On the other hand, the study of CSIs allows us to study the values of the variables that really discriminate among dependence and independence structure by neglecting the unnecessary parameters. For this reason it is possible to develop strategies concerning the values of the conditioning variable where the independence does not hold. Further developments on these models may regard the multivariate regression models associated, similarly to the approach of [8].

References

1. Bartolucci, F., Colombi, R., Forcina, A.: An extended class of marginal link functions for modelling contingency tables by equality and inequality constraints. Stat. Sin. **17**, 691–711 (2007)
2. Boutilier, C.: Context-specific independence in Bayesian networks. In: Proceedings of the Twelfth International Conference on Uncertainty in Artificial Intelligence. Morgan Kaufmann Publishers Inc., Burlington (1996)
3. Cazzaro, M., Colombi, R.: Marginal nested interactions for contingency tables. Commun. Stat. Theory Methods **43**(13), 2799–2814 (2014)
4. Colombi, R., Giordano, S., Cazzaro, M.: hmmm: an R package for hierarchical multinomial marginal models. J. Stat. Softw. **59**(11), 1–25 (2014)
5. Drton, M.: Discrete chain graph models. Bernoulli **15**(3), 736–753 (2009)
6. ISTAT: Italian innovation survey 2002–2012 (2015). https://www.istat.it/it/archivio/87533
7. Lauritzen, S., Wermuth, N.: Graphical models for associations between variables, some of which are qualitative and some quantitative. Ann. Stat. **17**(1), 31–57 (1989)
8. Marchetti, G., Lupparelli, M.: Chain graph models of multivariate regression type for categorical data. Bernoulli **17**(3), 827–844 (2011)
9. Nicolussi, F.: Marginal parametrization for conditional independence models and graphical models for categorical data. Ph.D. thesis (2013)
10. Nicolussi, F., Cazzaro, M.: Context-specific independencies for ordinal variables in chain regression models. arXiv:1712.05229 (2017)
11. Nyman, H., Pensar, J., Koski, T., Corander, J.: Context-specific independence in graphical log-linear models. Comput. Stat. **31**(4), 1493–1512 (2016)
12. Nyman, H., Pensar, J., Corander, J.: Context-specific and local independence in Markovian dependence structures. In: Dependence Logic, pp. 219–234. Springer, Basel (2016)
13. Roverato, A., La Rocca, L.: Log-mean linear models for binary data. Biometrika **100**(1), 485–494 (2013)
14. Rudas, T., Bergsma, W.P., Németh, R.: Marginal log-linear parameterization of conditional independence models. Biometrika **97**(4), 1006–1012 (2010)

Part V
Big Data Analysis

Big Data and Network Analysis: A Combined Approach to Model Online News

Giovanni Giuffrida, Simona Gozzo, Francesco Mazzeo Rinaldi, and Venera Tomaselli

Abstract In recent years, large volumes of data are generated by automatic extraction of information, innovative data mining, and predictive analytics. This paper proposes an innovative approach by combining Big Data with the analysis of relational structures in order to improve actionable analytics-driven decision patterns. From the website of one of the largest online Italian newspapers, interactions among users and their comments about a 2016 Italian constitutional review bill are organized in a Big Data audience model. Readers' sentiments are measured and relational patterns are classified by descriptive measurements and clustering structures implemented in Network Analysis methods.

Keywords Big Data · Network Analysis · Relational patterns · Clustering structures

1 Big Data and Network Analysis

Today, in the Big Data (BD) world, the extraction of meaningful knowledge from the available data represents a severe challenge for data analysts. Data is "big" because of its high dimensionality, complexity, and heterogeneity: for example, textual or graphical data extracted from social networks are often mixed with socioeconomic data for more tailored marketing activities.

G. Giuffrida · S. Gozzo · V. Tomaselli (✉)
Department of Political and Social Sciences, University of Catania, Catania, Italy
e-mail: ggiuffrida@dmi.unict.it; simona.gozzo@unict.it; tomavene@unict.it

F. Mazzeo Rinaldi
Department of Political and Social Sciences, University of Catania, Catania, Italy

KTH, Royal Institute of Technology, School of Architecture and the Built Environment, Stockholm, Sweden
e-mail: fmazzeo@unict.it

F. Greselin et al. (eds.), *Statistical Learning of Complex Data*,
Studies in Classification, Data Analysis, and Knowledge Organization,
https://doi.org/10.1007/978-3-030-21140-0_19

Unfortunately, today traditional analytical tools quite often *break* when dealing with BD. As a matter of fact, due to the need of simultaneous processing, many statistical analytical techniques do not scale to BD. Suitable statistical techniques for dimensionality reduction are now needed for both data visualization and their numerical treatment. Specifically, to analyse BD sentiment analysis and classification methods are very useful.

Social media are a precious source of data to study public opinion and definition of new trends, offering insights into attitudes, behaviours, discourses, and social linkages among individuals. Social media data could be integrated with common data sources like online searches, text mining, mobile phone devices, sensors, satellite images, or Global Positioning Systems (GPS) for a more complete data repository to analyse.

The growing importance of social media data to be used in the social sciences arises from the availability and affordability of swiftly collected real-time large-scale data. With this data, we can follow individuals and their networks over time and across spaces, offering a new source of information, particularly when data quality is either poor or unavailable all together.

Digital social interactions generate data in large scale and at low cost. Reported events or surveyed opinions are sources of data about context, content, and meaning of social interactions [2]. Network Analysis (NA) methodology is the theoretical foundation approach to analyse networks depicting social interactions among actors [5]. Specifically, it is also employed in relational and structural analysis of decision-making [6] and organizational behaviour patterns [3]. NA methods measure structural attributes of networks—by *size, density, clustering, openness, stability, reachability, centrality* measurements [11]—and classify structural patterns.

In literature, we found one study about a large-scale analysis of the news media coverage of the 2012 US presidential elections campaign [10] that proposes a BD approach combined with NA methods. Key actors and their relations are extracted from the online media narrative of the US elections. And this information is organized as a directed semantic graph.

Taking into account the interest for understanding and improving actionable analytics-driven decision patterns, here we present a study based on a very large database generated by one of the largest online Italian newspapers. All the comments of the readers concerning news and articles about the Italian Senate constitutional reform are employed as sources of info. Focusing on the audience analysis system to detect online news, the findings of the study show how NA methods and BD tools can efficiently be integrated in the analysis of relational patterns to improve decision-making processes knowledge.

The approach is innovative in order to gain insights in the linguistic analysis of texts by extracting relational data. Combining BD mining techniques with NA measurements, the study aims to collect and analyse information about online readers to detect proxies of structural and positional network indicators. With this in mind, the network measurements are useful to define reliable impact indicators suitable to detect opinions and moods in real-time in terms of relevance, visibility, reachability, marginality, and resilience of BD.

We firstly developed an audience BD model to collect and manage info from users' interactions with published news and comments about a recent Italian constitutional review bill with important political implications. This is achieved by using the advanced search feature of one of the most diffused Italian online newspapers to extract text entities in order to profile users' targets linked to the interest and mood by reads and comments.

Afterwards, NA methods are employed to analyse the BD model measuring the relational structure of readers browsing on the newspaper website, focusing on the importance of selected news with a great impact in terms of capability to connect readers and commenters as web users in a net. Focusing on the audience analysis system to detect online news, the findings of the study show how NA methods and BD tools can efficiently be integrated in the analysis of relational patterns to improve the knowledge in the decision-making processes.

2 The Big Data Audience Model

Our BD audience model, already applied to other online newspapers [7], is a standard relational model composed of tables and relationships among those [4]. In a typical relational model, "tables" (i.e., "relations") are collections of "rows", where each row contains several fields arranged in columns. Two tables may be "joined" based on some "key" values. A "join" is a relationship between two tables and it can be of different types: "one to many", "many to many", or "one to one". "A one to many" relationship happens when one row in one table corresponds to many rows in the second table. "A many to many" relationship happens when many rows from one table correspond to many rows in the second table. Specifically, in our model, a user performs many actions (i.e., reads many articles on the newspaper), giving rise to "a one to many" relationships between the "user" table and the "action" table.

In our BD audience model, each user browses the website and performs actions towards specific contents. A "content" is a general *html* document published on the newspaper website. An "action" is a pageview generated by a user on a given content. Using the advanced search feature available on the newspaper site, we retrieved all articles containing the "riforma Senato" string in the time period from January 1st, 2014 through December 31st, 2014. Thus, we collected all 47 articles.

After a manual check, we discarded 29 of those as they were not actually focused on the specific topic but on other political issues. The result was 18 articles from the "Politics" section of the newspaper about the Senate reform bill, published in the time frame from March 12th, 2014 until August 8th, 2014. In the same time frame, collectively, in the "Politics" section of that newspaper, a total of 1788 articles were published. After an article is published on the newspaper, readers can place comments on that article; this basically starts a blog around that specific article. We assume that the larger the number of comments to a specific article, the more that article raised readers' interest. Thus we use the number of comments to each article as proxy for the readers' interest to that article. The 18 specific articles we selected generated a total of 886,898 pageviews and 2461 comments. Whereas the 1788 general politics articles generated a total of 32,774,270 pageviews and 14,108 comments. In order to compare the readers' interest for the specific Senate reform articles, we computed some quantitative measures and specifically:

- *the average number of pageviews per article*. This measures readers' interest on the topic: 49,272 for the 18 articles versus 18,330 for the 1788 general articles.
- *the average number of comments per article*. This measures readers' engagement with the topic: 137 for the specific articles versus 8 for the general articles.
- *the probability of generating new comments*. This is the probability that a reader writes a new comment after reading the article: 0.28% versus 0.04% for specific and general articles, respectively.

These numbers clearly show a significant interest of the readers for the Senate reform. Furthermore, in order to check the relevance of the Senate reform, we compare NA results obtained from two data samples: the first includes 18 articles about the Senate reform, while the second one 18 random articles selected from political section of the newspaper.

Our analysis focuses on the measurement of a readers' mood (from negative to positive) for each comment. Thus, we label each comment through a sentiment analysis algorithm with a score ranging from -1 to $+1$. Both the sentiment analysis and the extraction of semantic entities contained in the 18 articles were made using a public Application Programming Interface (A.P.I.) service for Semantic Text Analytics, provided by https://dandelion.eu/. We found out that the entities that received most of the comments were: "Senato", "Governo", "Parlamento", "Senatore", and "Matteo Renzi", with an average sentiment score of -0.426. Since the comments in a newspaper generally tend to be negative (see, among others, [9]), this negative sentiment score does not immediately imply that commenters were against the proposed bill. Comparing the average sentiment score of the top entities with the average sentiment score of the comments in the general Politics section (-0.38), the commenters have a slightly positive sentiment (0.04), representing only 8% of the standard deviation measured on comments in 0.5.

3 The Network Analysis Application

We employed NA methods to describe relational BD drawing in a network graph, points (nodes) and lines (edges) connecting them, by Gephi and NodeXL tools. In order to detect the main relational clusters, according to the NA criteria [11], the data structure is modified assuming that the nodes are the reference objects and the links among nodes are the units of analysis. For this purpose the analysis does not focus on readers' attributes but on the importance of news as their capability to connect people according to the following rules:

- two nodes/users are connected if they read the same articles also at different times,
- in the two graphs (Fig. 1) links between readers are represented by removing repetitive and recursive ties.

According with these rules, data from the two samples described in Sect. 2 are used to reconstruct each relational structure. Finally, two graphs are obtained: the first sample includes 1892 nodes and 386,952 edges among all registered readers visualizing 18 political articles (selected randomly); the second sample includes 15,550 nodes and 515,739 edges among nodes of the specific articles from the "Politics" section of the newspaper about the Senate reform bill.

With the aim to analyse the structural behaviour of graphs, the modularity measure and the clustering coefficient are employed. The modularity quantifies the quality of the division into "modules" or communities. A good subdivision has high values of modularity. The logic employed provides that the density of bonds will be high within communities and limited between communities. The modularity coefficient is a measure that shows the quality of communication in the network. The modularity index range can vary from 0 to 1 and a greater coefficient indicates the tendency of the whole network to be structured in cohesive groups with high internal connections but not very connected to each other. It is a good tool for our analysis goals [1, 8]. Specifically, this index identifies which and how many "communities" are formed on specific aggregates of information or if the interest for the issue (in our case, the Senate reform) is, instead, "generic" or "generalized". This can happen when there are no different "communities" but the network is defined rather as unitary.

The graph modularity is 0.37 for links among political news and 0.18 for news about the Senate reform, then modularity shows that the graph (Fig. 1) on Senate reform is more "compact" compared to the graph about general political issues. This latter shows separated subgroups and a weak community structure. However, both graphs have the same clustering coefficient (0.84), where this index measures the level at which the nodes of a graph tend to be connected through the proportion of closed triple bonds within the network. The clustering coefficient measures, in particular, how much the nodes to which a focal node is linked are in turn connected to each other.

In this study, this coefficient can be particularly useful for evaluating the importance attributed to some news or information, evidently so "interesting" to produce a convergence of interests in little groups. In fact, the measure takes into account the connections focused on specific issues. Since both graphs in Fig. 1 have the same clustering index but different values of the modularity coefficients, this suggests that the focus could be—in this case—on the identification of problems and themes that probably include different dimensions and arguments rather than single news or information which polarize the attention of specific, small groups.

The Clauset–Newman–Moore algorithm [1] permits to point out cohesive clusters within the graph, grouping the vertices by means of modularity. This is a method to select more cohesive groups within the net. It is based on modularity as property of the network specifying division of network into "communities". The community structure implies to group vertices if there is a higher density of edges within groups than between them. The problem of detecting such communities within networks has been recently studied. We choose the Clauset–Newman–Moore algorithm because it works well using sophisticated data structures and extremely large networks. The first graph (with links about political articles) shows intersections among seven collapsed groups, while the second (Senate reform) four clusters (Fig. 1).

Here nodes are not singular web users but overlapped cliques (the groups are all nodes strongly connected each other). However, each cluster has a specific relational structure and presents specific traits if compared to the structural characteristics of the network. Graph about general political news is (as expected) more heterogeneous. Finally, we obtain seven different subgroups. These clusters emerge because the links among nodes are higher compared to that detectable in the overall structure. Besides, there are strong structural differences among groups (Table 1). This means that there are seven different profiles of users connected from a similar selection of news visualized.

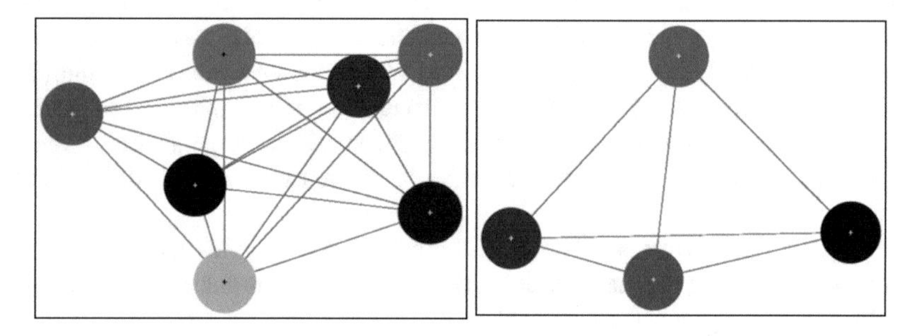

Fig. 1 Collapsed seven groups on general political news (on the left side) and four groups on Senate reform (on the right side)

Table 1 Degree, betweenness, and clustering (average values) for groups on political news issue

Political news	Degree	Betweenness	Clustering
Tot (averages)	379.04	756.27	0.84
Group 1	402.25	1353.45	0.73
Group 2	355.98	371.68	0.91
Group 3	484.29	1063.77	0.79
Group 4	274.57	102.57	0.96
Group 5	315.26	824.47	0.85
Group 6	406.00	0.00	1.00
Group 7	252.63	115.68	0.96

Table 2 Degree, betweenness, clustering (average values) for groups on senate reform issue

Senate reforms	Degree	Betweenness	Clustering
Tot (averages)	517.21	515.99	0.85
Group 1	411.98	748.35	0.81
Group 2	605.00	184.02	0.91
Group 3	626.41	568.31	0.80
Group 4	412.20	113.50	0.98

First, second, and third groups are the most important, with both high degree (more articles viewed) and betweenness (more brokers). The centrality measure distinguishing the most relevant subgroups is the betweenness for the first graph and the degree for the second (comparing averages in Tables 1 and 2). It is important to notice that by selecting a more specific section of the whole graph is like applying a "magnifying glass" on the whole structure. Zooming-in, we notice more clearly roles and information originally hidden in the graph. As a matter of fact, the intermediary function between "groups" is now visible through the centrality betweenness measure, while it was originally hidden by the strong overlap between multiple groups.

NA tools can be useful to drive the decision-making process since they show the relational structure among web users. The network tools, indeed, allow detecting which news lead to more or less homogeneous groups. The integration of this information with others, which also take into account comments and relational profiles, can be particularly useful to drive decision-making process. It is also possible to carry out analyses that refer to groups and not to single subjects as a reference unit. In this way, links within the groups are more or less dense according to the common incidence attributed to specific information. This behaviour cannot be observed by selecting only individual elements. One image we obtain is referred to the diameter and the shortest paths in the graph. This information is about the number or the kind of items that connect all people within the net. Another information is about the identification of central or peripheral subgroups compared to the whole net.

This relational structure could mean that there are groups of readers who intercept multiple cross information (there is not a specific kind of information but these readers simply select the more important news from time to time). This is the main feature of the first group, where the most informed nodes, with the highest betweenness measure, are the "attractors" of the network, the structure of the graph depending by them.

The configuration indicates that—within these groups—there are readers with a greater heterogeneity in cultural consumption, able to "connect" with very different contexts and arguments, acting as a structural "broker" among other (more selective) groups. Groups of this type have a thematically more varied configuration and they are more likely to consist of an average informed user, who will tend to select the main information on lines differentiated issues.

Instead, a network characterized by subgroups with higher degree centrality (as the second graph) is more homogeneous with respect to the selected items. In this case, few issues convey the interest. The highest degree value means that nodes are directly and strongly connected each other (see often the same news): readers are more specialized.

4 Conclusions

We proposed an innovative blended approach between BD mining and NA methods, useful to identify positional (central *vs* peripheral) and relational (brokers *vs* isolated) structures of nodes or units within BD. We modelled a very large dataset containing all readers' comments on articles published by a prominent Italian newspaper. We focused on the political debate around a proposed reform for a change of a constitutional bill. The dataset at hand, besides being very large in nature, is collected in real-time, and this offers another interesting dimension for data analysis. This may turn into a very valuable tool for policy makers in general as they can measure, in real-time, readers' reaction to new law proposals. As a consequence, such a data processing can lead to very meaningful results both to measure overall readers' sentiment and support the decision-making process.

In our work, we mostly focused on users' comments collected by the online newspaper. In the future, we will try to extend the dataset by including additional data such as user's demographic (if available for the registered users) and behavioural data, that is, data about interactions over time between each user and the newspapers contents. We believe that this additional type of data could bring some very valuable insights when combined with the commenting activity analysed in the present work. And this should lead us to better and more refined models.

References

1. Clauset, A., Newman, M. E., Moore, C.: Finding community structure in very large networks. Phys. Rev. E, **70**, 66–111 (2004)
2. Cronin, B.: A window on emergent European social network analysis. Procedia Soc. Behav. Sci. **10**, 1–4 (2011)
3. Cross, R., Parker, A.: The Hidden Power of Social Networks: Understanding How Work Really Gets Done in Organizations. Harvard Business School Press, Boston (2004)
4. Date, C.J.: An Introduction to Database Systems, pp. 592–597. Addison Wesley, Boston (2004). ISBN 978-0-321-19784-9
5. De Nooy, W.: Structure from interactive events. BD&S (2015). https://doi.org/10.1177/2053951715603732
6. Laumann, E.O., Pappi, F.U.: Networks of Collective Action: A Perspective on Community Influence Systems. Academic Press, New York (1976)

7. Mazzeo Rinaldi, F., Giuffrida, G., Negrete, T.: Real-time monitoring and evaluation. Emerging news as predictive process using Big Data based approach. In: Petersson, G.K., Breul, J.D. (eds.) Cyber Society, Big Data, and Evaluation. Series: Comparative Policy Evaluation, vol. 24, Transaction, New Brunswick, NJ (2017)
8. Newman, M.E.J., Girvan, M.: Finding and evaluating community structure in networks. Phys. Rev. E **69**, 26–113 (2004)
9. Reis, J., Benevenuto, F., Olmo, P., Prates, R., Kwak, H., An, J.: Breaking the News: First Impressions Matter on Online News. In: Proceedings of the Ninth International AAAI Conference on Web and Social Media, pp. 357–366. Oxford University, Oxford (2015)
10. Sudhahar, S., Veltri, G.A., Cristianini, N.: Automated analysis of the US presidential elections using Big Data and network analysis. BD&S (2015). https://doi.org/10.1177/2053951715572916
11. Wasserman, S., Faust, K.: Social Network Analysis: Methods and Applications. Cambridge University Press, Cambridge (1994)

Experimental Design Issues in Big Data: The Question of Bias

Elena Pesce, Eva Riccomagno, and Henry P. Wynn

Abstract Data can be collected in scientific studies via a controlled experiment or passive observation. Big data is often collected in a passive way, e.g. from social media. In studies of causation great efforts are made to guard against bias and hidden confounders or feedback which can destroy the identification of causation by corrupting or omitting counterfactuals (controls). Various solutions of these problems are discussed, including randomisation.

Keywords Big data · Model bias · Experimental design · Nash equilibrium

1 The Challenges of Experimental Design with Big Data

The value of experimental design in physical and socio-medical fields is increasingly realised, but at the same time systems under consideration are more complex. It may not be possible to do a carefully controlled experiment in many areas, but at the same time huge quantities of data are being produced, for example, from social media and web-based transactions. An added problem is that the traditions of experimental design differ. For example, in engineering design it will be possible to do a control experiment on a test bench, whereas in the social-medical sciences the local counterfactual will be missing: we do not know how a particular patient would have fared if they were not given the drug. Foundation work on these issues is by Rosenbaum and Rubin [11]. Roughly, the causal effect can only be measured on the average, with great care taken about the background population, with more reluctance than in the physical sciences to extend the conclusions outside the

E. Pesce (✉) · E. Riccomagno
Department of Mathematics, University of Genoa, Genova, Italy
e-mail: pesce@dima.unige.it; riccomagno@dima.unige.it

H. P. Wynn
Department of Statistics, London School of Economics, London, UK
e-mail: h.wynn@lse.ac.uk

F. Greselin et al. (eds.), *Statistical Learning of Complex Data*,
Studies in Classification, Data Analysis, and Knowledge Organization,
https://doi.org/10.1007/978-3-030-21140-0_20

population under study. An old issue, which goes back into the history of science, is the distinction between active and passive observation. Is placing a sensor on a driverless car to collect data (for control) an intervention in the sense of the declaration that to prove causation you have to intervene? Despite these different historical traditions there seems to be general agreement (1) that deriving causal models is a kind of gold standard and (2) that to produce a causal model we need to guard against bias from different sources: hidden confounders, sampling bias, incomplete models, feedback and so on.

We cover a few of the ideas from the theory of causation (Sect. 2) and then suggest that the double activity of building causal models while at the same time guarding against bias has features of a cooperative game (Sect. 3.1). At its simplest a randomised clinical trial is minimax solution to a game against the sources of bias. With this in mind we make the natural but speculative suggestion that we can import theories of Nash equilibrium and supply a simple example motivated by the theory of optimum experimental design under a heading of optimal bias design. We could have taken a Bayesian optimal design, for example, from [7, 13]. But for this short paper we felt it was enough to allow our randomness to come from the error distribution or the randomisation itself.

2 Causal Models

A major critique of passive analysis of the machine-learning type is the lack of attention to the building of causal models. We discuss briefly the main ingredients of causal graphical models and then the implications for experimental design [12].

A causal model is often described via a direct acyclic graph (DAG), $G(E, V)$, where each vertex $i \in V$ holds a (possibly vector) random variable X_i. Care has to be taken with the edges $i \rightarrow j$. The natural intuition that the edge means X_i causes X_j is not correct, at least not without much qualification. The DAG is a vehicle for describing all conditional independence structures.

We can define a variable X_j which is never observed as *latent*, also *hidden*. There is a slight difference: hidden may be that we do not know it is there but it might be. Latent may also be taken as expressing prior information. Thus a latent layer in machine-learning context may be included to allow a more complex model, such as a mixture model.

The conundrum with causal models stems from the distinction between passive observation and active experimental design. Experimental design is an *intervention* and there are essentially two types. First, we can simply apply some kind of *treatment* at node i to obtain a special X_i, for example, give a patient i a drug. Second, and even more active, one can set variable X_i to say high and low levels.

Passive observation means that a joint sampling distribution covers all observed X_i. The act of setting should be thought as advantageous in the sense that we are in some kind of classical or optimal design framework, but disadvantageous in that it is destructive. Roughly, setting X_i destroys our ability to learn about the population from which X_i comes.

Consider a simple DAG: $X_1 \longrightarrow X_2 \longrightarrow X_3 \longrightarrow X_4$ and for ease of explanation we write down a univariate linear version with obvious interpretation

$$X_1 = \theta_0 + \epsilon_1 \qquad X_2 = \theta_1 X_1 + \epsilon_2$$
$$X_3 = \theta_2 X_2 + \epsilon_3 \qquad X_4 = \theta_3 X_3 + \epsilon_4,$$

where $\{\epsilon_i\}_{i=1,\dots,4}$ are error variables. Suppose we are interested in the last causal parameter θ_3. Ideal would be to carry out a controlled experiment, setting the levels of X_3 and observing X_4. The first assumption to make is governed by the following:

Principal 1 The distribution of X_4 conditional on a *set* value of X_3 is the same as when the same value of X_3 was passively observed.

There are arguments to justify this but it remains a most important assumption. We can also passively observe X_1, X_2, X_3, X_4. Note that the model is *nonlinear* in the parameters as $X_4 = \theta_0\theta_1\theta_2\theta_3 + \theta_1\theta_2\theta_3\epsilon_1 + \theta_2\theta_3\epsilon_2 + \theta_3\epsilon_3 + \epsilon_4$ and also that X_4 is Gaussian if the $\{\epsilon_i\}$ are Gaussian. One may not have to choose between a controlled experiment and passive observation. This leads to another principal, see [6].

Principal 2 A mixture of passive observation experiment and active experimentation may be optimal.

There is considerable discussion in trying to understand how to learn for DAG models with interventions, and controlled experiments are a form of intervention. Most effort has been put into identifiability; see [4] for a review. In our example suppose there is an extra arrow $X_1 \rightarrow X_4$. Such an arrow is referred to as a backdoor. If the index is time we can say that there is another path from X_1 into the future in addition to $X_1 \rightarrow X_2 \rightarrow X_3 \rightarrow X_4$.

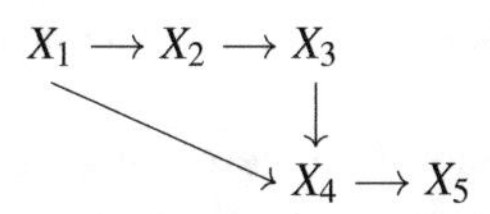

Now if we fix X_3 we cannot so simply estimate θ_3 because the distribution of X_4 is corrupted by the new path. In the observational case, we have another parameter and the changed equation

$$X_4 = \theta_3 X_3 + \theta_4 X_1 + \epsilon_4. \tag{1}$$

There are now too many parameters for the observations (even with replication).

The celebrated backdoor theorem due to [10] tells us how to obtain identifiability. Suppose you want to see whether X_i causes X_j, then we need two conditions for a good conditioning set of variables S:

1. No node (variable) in S is a descendent of X_i.
2. S blocks every (backdoor) path from X_i to X_j that has an arrow into X_i.

This theorem tell us: (1) whether there is confounding given this DAG, (2) if it is possible to remove the confounding and (3) which variables to condition on to eliminate the confounding. For example, if we are trying to establish the effect of X_3 on X_4, then we must observe, or set and condition on, any X_i which is not a descendant of X_3 and blocks all paths from, in our case, ancestors of X_3. In addition, if there are any other downstream (future) variables such as an extra X_5 with $X_4 \rightarrow X_5$, then X_5 will not interfere with our causal analysis; we can forget it. In summary

Principal 3 Guard against effects from nuisance confounders by suitable additional conditioning.

3 Bias Models

Before presenting our contribution, we briefly review relevant literature. For the model without bias

$$\mathrm{E}\left[Y_i\right] = \theta^T f(x_i) \tag{2}$$

with $\theta \in \mathbb{R}^p$, $x_i \in \mathscr{X}$ for $i = 1, \ldots, n$, and the usual assumptions on the random error, [3] and [15] propose information-based and sequential algorithms (also response adaptive in [3]) for the selection of a subsample from a large, or possibly big, dataset. They provide an optimal subsample with respect to a chosen utility function.

Bias model and optimal design of experiments were considered by Montepiedra and Fedorov [9] and recently in the context of big data by Wiens [16]. Those authors add a bias term $\delta^T h$ to the model (2) and thus study $\mathrm{E}\left[Y_i\right] = \theta^T f(x_i) + \delta^T h(x_i)$. They search for a design which minimises the mean square error of the least square estimator of the θ parameters, guarding θ from the bias term. In particular [16] proposes a theory of minimax I- and D-robust design as subset of a large finite set of points, while [9] proves results for a design to be optimal when the effect of the bias term is bounded above from a given constant and below from zero.

The conditioning argument of the backdoor theorem is a way of avoiding biases. In the above example in Eq. (1) θ_4 gives a bias. Enough conditioning creates a kind of laboratory inside which we can conduct our experiment by setting the level of X_3. Sometimes this is referred to as creating a Markov blanket. But there are sources of bias which either we do not know at all or have some ideas about but are too costly to control. Biases range from those we really know about but simply do not observe to those which are introduced to model additional variability. This will affect the overall distribution of the observed variables, in a way similar to classical factor analysis.

Principal 4 Special models are needed to ascertain and guard against hidden sources of bias, for example, using randomisation or latent variable methods.

We build on the ideas in [9] and discuss in detail how optimal experimental design can guard against hidden sources of bias, indicated below with the letter z. Thus consider a two part model in which the first part is the causal model of main interest with parameters θ and the second part is the bias term with parameters ϕ. This separation is familiar from traditional experimental design where θ and ϕ might be treatment and block parameters, respectively [1, 9]. The model is

$$Y_i = \theta^T f(x_i) + \phi^T g(z_i) + \epsilon_i, \tag{3}$$

where the ϵ_i are independent and have equal variance σ^2.

We want to protect the usual least square estimator, $\hat{\theta}$, obtained from the reduced model in Eq. (2) ignoring the bias term $\phi^T g(z_i)$. Define the full moment matrix by

$$M = \int (f(x)^T, g(z)^T)^T (f(x)^T, g(z)^T) \xi_{x,z}\, d(x, z) = \begin{bmatrix} M_{11} & M_{12} \\ M_{21} & M_{22} \end{bmatrix},$$

where $\xi_{x,z}$ is the experimental design measure over (x, z)-space. Then the mean squared error (MSE) matrix can be written as

$$\mathbb{E}\{(\hat{\theta} - \theta)(\hat{\theta} - \theta)^T\} = \sigma^2 N^{-1} R,$$

where

$$R = M_{11}^{-1} + M_{11}^{-1} M_{12} \psi \psi^T M_{21} M_{11}^{-1} = S_1 + S_2$$

with $\psi = \frac{N}{\sigma}\phi$ the standardised bias parameter and N the sample size (see [9]).

Well-known criteria for optimality ask to minimise over the choice of experimental design the quantity: trace(R) = trace(S_1) + trace(S_2) (the trace criteria or A-optimality) or $\det(R) = \det(S_1)\left(1 + \psi^T M_{21} M_{11}^{-1} M_{12} \psi\right)$ (the D-optimality criteria).

The design problem is easier when the design space and design D are direct products and thus can be written as

$$\mathscr{X} \times \mathscr{Z}, \quad D = D_1 \times D_2 \tag{4}$$

with $x \in \mathscr{X}$ and $z \in \mathscr{Z}$, D_1 and D_2 are finite subsets of, respectively, $\mathscr{X}$ and $\mathscr{Z}$. Then, trace(R) includes a term which depends only on D_1, likewise a factor in $\det(R)$ depends only on D_1.

The most familiar example is from clinical trials where one compares a treatment against a control. Consider the simple case

$$Y_{1i} = \theta_1 + \theta_2 + \phi(z_{1i} - \bar{z}) + \epsilon_i$$
$$Y_{2j} = \theta_1 - \theta_2 + \phi(z_{2j} - \bar{z}) + \epsilon_j,$$

where the z_i are unwanted confounders which may be a source of bias, the $\bar{z}$ is the grand mean and $n = N/2$ points are allocated to each group. Adapting the above analysis we obtain

$$M = \frac{X'X}{N} = \begin{bmatrix} 1 & 0 & 0 \\ 0 & 1 & (\bar{z}_1 - \bar{z}_2)/2 \\ 0 & (\bar{z}_1 - \bar{z}_2)/2 & s \end{bmatrix},$$

where the $\bar{z}_i$, $i = 1, 2$, terms are the group means and $Ns = \sum_{i=1}^{n}(z_{1i} - \bar{z})^2 + \sum_{i=1}^{n}(z_{2i} - \bar{z})^2$. The bias term is trace$(S_2) = \psi^2(\bar{z}_1 - \bar{z}_2)^2/4$ which is zero when $\bar{z}_1 = \bar{z}_2$. This is the simplest case of balance and extends easily to multivariate z. A number of methods of achieving balance have been studied, each of which can be cast in the above framework:

1. Stratification: balancing in each stratum and then aggregating the difference.
2. Distance methods: pairing up treatment and control with which are close in z-space with respect to some distance such as Mahalanobis distance [8].
3. Propensity score. This much researched method seeks to balance in such way as to ensure that the bias correction is extended to a larger parent population [11, 12]. Some adaptation of the above method analysis is possible in this case.

3.1 A Game Theoretic Approach

For ease of explanation we introduce two players: Alice (A) and Bob (B). Alice selects a causal model design D_1 using $\{\theta, f\}$ and Bob selects design D_2 using $\{\phi, g\}$. In the product case (4), Alice and Bob can operate separately. In other cases they may cooperate fully to find the best design over the design space for the pair (x, z). However there is another possibility, namely to use a Nash equilibrium approach [2, 5, 14].

For two players A and B with composite cost functions $C_1(u, v)$, $C_2(u, v)$ and solutions u^*, v^* at equilibrium it holds

$$\text{Alice}: u^* = \underset{u}{\text{argmin}}\ C_1(u, v^*)$$
$$\text{Bob}: \ v^* = \underset{v}{\text{argmin}}\ C_2(u^*, v).$$

We illustrate the presence of Nash equilibrium in causation-bias setup by a simple example. We take a distorted design space, but still a product-type design measure. Thus, let the model be

$$\mathbb{E}(Y) = \theta_0 + \theta_1 x + \phi z$$

and let the design have a four support points (we put the design measure in the second line):

$$\left\{ \begin{array}{cccc} (1,1), & (0,1), & (0,-1), & (-1,-1) \\ \alpha\beta, & (1-\alpha)\beta, & \alpha(1-\beta), & (1-\alpha)(1-\beta) \end{array} \right\},$$

where $0 \leq \alpha, \beta \leq 1$

Since, in this case, M_{12} is a 2×1 column vector:

$$\text{trace}(S_2) = \psi^2 M_{21} M_{11}^{-2} M_{12}.$$

The equilibrium takes the form:

$$\begin{aligned} \text{Alice}: \alpha^* &= \arg\min_\alpha \text{trace}(S_1) \\ \text{Bob}: \ \beta^* &= \arg\min_\beta \text{trace}(S_2). \end{aligned}$$

There are two Nash equilibria given by solving $\frac{\partial}{\partial\alpha}\text{trace}(S_1) = \frac{\partial}{\partial\beta}\text{trace}(S_2) = 0$. This gives two solutions (α^*, β^*) and $(\frac{1}{2}, \frac{1}{2})$ with $\alpha^* = 0.59306$ and $\beta^* = 0.08274$ computed numerically. Note that both solutions do not depend on ψ, and in fact scale invariance of this kind is a well-known feature of Nash equilibrium.

We can compare the solution with an overall optimisation by setting $\psi = 1$ and minimising $\text{trace}(S_1) + \text{trace}(S_2)$. The minimum is 4, it is achieved at $(\alpha, \beta) = (\frac{1}{2}, \frac{1}{2})$ with $(\text{trace}(S_1), \text{trace}(S_2)) = (3, 1)$. Whereas at (α^*, β^*) the value of $\text{trace}(S_1) + \text{trace}(S_2)$ is approximated to 5.1735 with $(\text{trace}(S_1), \text{trace}(S_2)) = (4.483, 0.6905)$.

Let us return to the role of Bob in our narrative. His experimental design decision will depend on his knowledge about the bias. For ease of explanation we reduce the argument to two canonical cases.

Approach 1 Unknown ψ

$$\text{trace}(S_2) = \text{trace}\left(M_{11}^{-1} M_{12} \psi \psi^T M_{21} M_{11}^{-1}\right) = \psi^T Q_1 \psi$$

$$Q_1 = M_{21} M_{11}^{-2} M_{12}.$$

Under a restriction $||\psi|| = 1$ this achieves a maximum at the maximum eigenvalue: $\lambda_{\max}(Q_1)$. We can take this as our criterion which is close to the E-optimality of optimum design theory.

Approach 2 In Eq. (3), for unknown $\phi^T g(z) = h(z) \in \mathcal{H}$ in some function class, we have

$$||\mathrm{E}(\hat{\theta}) - \theta||^2 = h(z)^T Q_2 \, h(z)$$

$$Q_2 = X_1 M_{11}^{-2} X_1^T.$$

where $X_1 = [f(x)]_{x \in D_1}$. We cannot optimise over x (X_1) because, in our narrative, Alice needs it for the causal parameter θ. A solution is then

$$\min_{P_z} \mathbb{E}_Z \left\{ \max_{h \in \mathcal{H}} \left(h(z)^T Q_2 \, h(z) \right) \right\},$$

where P_z is the randomisation distribution. In the language of game theory this is a mixed strategy to achieve a minimax solution.

Randomisation has been heralded as one of the most important contributions of statistics to scientific discovery. There are several arguments put forward for using randomisation: (1) it helps support assumptions of exchangeability in a Bayesian analysis, (2) it supports classical zero mean and equal variance arguments and (3) it produces roughly balanced samples.

4 Conclusion

After a discussion of some issues related to the use of experimental design to help establish causation in complex models, we study in a little more detail the use of optimal design methods to remove bias. In the standard case the causal part of a model can be estimated orthogonally from the bias. In more complex cases the problem can be set up as a cooperative game. We demonstrate the existence of Nash equilibria for a small example and point to a formulation which would include randomisation. This is a preliminary work, establishing model classes (for example, special h's, $\mathcal{H}$'s, P_z's) and conditions on D for which Approaches 1 and 2 can be turned into efficient algorithms is object of current work. The general proposition is that such methods will help protect causal models against bias.

Acknowledgements We thank the anonymous reviewers for thorough reading of the manuscript.

References

1. Box, G.E., Draper, N.R.: A basis for the selection of a response surface design. J. Am. Stat. Assoc. **54**(287), 622–654 (1959)
2. Cheng, C.S., Li, K.C.: A minimax approach to sample surveys. Ann. Stat. **11**, 552–563 (1983)

3. Drovandi, C.C., Holmes, C., McGree, J.M., Mengersen, K., Richardson, S., Ryan, E.G.: Principles of experimental design for big data analysis. Stat. Sci. **32**(3), 385–404 (2017)
4. Drton, M., Weihs, L.: Generic identifiability of linear structural equation models by ancestor decomposition. Scand. J. Stat. **43**(4), 1035–1045 (2016)
5. Grant, W.C., Anstrom, K.J.: Minimizing selection bias in randomized trials: A Nash equilibrium approach to optimal randomization. J. Econ. Behav. Organ. **66**(3), 606–624 (2008)
6. Hainy, M., Müller, W.G., Wynn, H.P.: Approximate Bayesian computation design (ABCD), an introduction. In: mODa 10–Advances in Model-Oriented Design and Analysis, pp. 135–143. Springer, Heidelberg (2013)
7. Hainy, M., Müller, W.G., Wynn, H.P.: Learning functions and approximate Bayesian computation design: ABCD. Entropy **16**(8), 4353–4374 (2014)
8. LaLonde, R.J.: Evaluating the econometric evaluations of training programs with experimental data. Am. Econ. Rev. **76**, 604–620 (1986)
9. Montepiedra, G., Fedorov, V.V.: Minimum bias designs with constraints. J. Stat. Plan. Infer. **63**(1), 97–111 (1997)
10. Pearl, J.: Causality. Cambridge University Press, Cambridge (2009)
11. Rosenbaum, P.R., Rubin, D.B.: The central role of the propensity score in observational studies for causal effects. Biometrika **70**(1), 41–55 (1983)
12. Rubin, D.B.: Bayesian inference for causal effects: the role of randomization. Ann. Stat. **6**, 34–58 (1978)
13. Sebastiani, P., Wynn, H.P.: Maximum entropy sampling and optimal Bayesian experimental design. J. R. Stat. Soc. B **62**(1), 145–157 (2000)
14. Stenger, H.: A minimax approach to randomization and estimation in survey sampling. Ann. Stat. **7**, 395–399 (1979)
15. Wang, H., Yang, M., Stufken, J.: Information-based optimal subdata selection for big data linear regression. J. Am. Stat. Assoc. **114**(525), 393–405 (2019)
16. Wiens, D.P.: I-robust and D-robust designs on a finite design space. Stat. Comput. **28**(2), 241–258 (2018)

Printed in the United States
By Bookmasters